System Development Charges

for

Water, Wastewater,

and

Stormwater Facilities

Arthur C. Nelson

LEWIS PUBLISHERS

Boca Raton   New York   London   Tokyo

Library of Congress Cataloging-in-Publication Data

Nelson, Arthur, C.
    System development charges for water, wastewater, and
stormwater facilities / author, Arthur C. Nelson.
        p.   cm.
    Includes bibliographical references and index.
    ISBN 1-56670-037-X
        1. Waterworks—United States. 2. Sewerage—United States.
    3. Impact fees—United States. I. Title.
    HD4461.H45   1995
    363.72′84′0973—dc20   94-23520
                                                                                    CIP

© 1995 by CRC Press, Inc.
Lewis Publishers is an imprint of CRC Press

No claim to original U.S. Government works
International Standard Book Number 1-56670-037-X
Library of Congress Card Number 94-23520
Printed in the United States of America 1 2 3 4 5 6 7 8 9 0
Printed on acid-free paper

Disclaimer

This book includes information directly, indirectly, or in compiled form from authentic sources. Reprinted copyrighted material is quoted with permission and sources are indicated. A wide variety of references are listed. Every reasonable effort has been made to give reliable data, information, and outcomes, but the author and the publisher cannot and do not assume responsibility for the validity of all materials and procedures or for the consequences of their use.

Dedication

To my long-time mentors and friends in this line of work,
Jim (James C.) Nicholas and Dick (J. Richard) Recht.

And to the hundreds of professionals over 3 decades from whom
I have learned much and for whom I am pleased to have served.

Author

Arthur C. Nelson, Ph.D., ASCE, ALCP, is a professor serving on the faculties of city planning, public policy, and international affairs at the Georgia Institute of Technology, where he is the coordinator of the graduate and undergraduate certificate programs in land development and in urban policy.

Dr. Nelson has more than 20 years of experience in comprehensive land use and facility planning, capital improvements programming, facility finance, and impact assessment. He has counseled more than 3000 local, state, federal, and foreign professionals, government agencies, and consulting firms in these and related areas. Through testimony and briefings, Dr. Nelson has advised city councils, county commissions, state public utility commissions, state legislatures, Congress, and ministries of foreign governments in these areas. In particular, Dr. Nelson has advised the National Science Foundation, National Academy of Sciences, U.S. Department of Commerce, U.S. Department of Housing and Urban Development, U.S. Department of Transportation, Her Majesty's Department of the Environment (United Kingdom), the Federal National Mortgage Association, the American Planning Association, the American Society of Civil Engineers, the Urban Land Institute, the National Association of Home Builders, and the Lincoln Institute of Land Policy.

The author received a B.S. in political science with minors in geography, urban studies, and social service from Portland State University in 1972. From 1972 through 1984, Dr. Nelson had his own Northwest-based consulting firm and continued his graduate studies at Portland State University. He earned a master of urban studies degree in public administration in 1976, and a doctorate of urban studies in regional science and regional planning in 1984. He served on the regional and community planning faculty at Kansas State University (1984–1985) and on the urban and regional studies faculty at the University of New Orleans (1985–1987) before joining the Georgia Tech faculties in 1987.

Dr. Nelson has more than 120 scholarly and professional publications in journals, proceedings, and books. His articles have appeared in such journals as *Land Economics, Journal of the American Planning Association, Land Use Policy, Growth and Change, Urban Studies, Journal of Urban Planning and Development,* and *Public Works.* He is author, co-author, or editor of: *Development Impact Fees: Policy Rationale, Theory, and Practice; Practitioner's Guide to Development Impact Fees; Estimating Land Use and Facility Needs and Impacts; Growth Management Principles and Practices;* and *The Regulated Landscape.*

Acknowledgments

This book is the product of many years' association with many fine professionals and scholars. I owe a special debt to Jim (James C.) Nicholas and Dick (J. Richard) Recht, from whom I learned much about this field of planning and public finance. Through Dick's California-based urban economics firm, Recht Hausrath and Associates (Oakland, CA), I learned much about and practiced facility capacity and financial analysis in the context of planning and environmental impact assessment. From Jim, professor of urban planning and affiliate professor of law at the University of Florida, Gainesville, I learned much about the application of legal and economic principles to the crafting of plans, policies, and programs to substantiate system development charges.

I owe an additional debt to hundreds of other professionals, but one is prominent. Jim Duncan and his Texas-based firm, James Duncan and Associates (Austin, TX), afforded me many opportunities to refine and apply my techniques throughout the southeastern, midwestern, and mountain United States.

Finally, I am indebted to my institution, the Georgia Institute of Technology, and its many fine administrators, faculty, and staff for affording me the opportunity to teach these principles to hundreds of professionals throughout the world. Special thanks to Thomas N. Debo, Steven P. French, and David S. Sawicki of the City Planning Program, and to Clifford R. Bragdon, W. Denny Freeston, and Kimberlee E. Rish of the Division of Continuing Education.

—

Contents

1   The Need for System Development Charges

INTRODUCTION

System development charges (SDCs) are one-time charges paid by new development to finance the construction of public facilities needed to serve it. SDCs represent a major departure from past local government facility financing policies. Historically, local residents and property owners have been charged taxes to build facilities to accommodate new growth, which in turn provides homes and jobs; but taxes have risen faster than taxpayers can tolerate. Moreover, the state and federal roles in financing water, wastewater, and stormwater systems have changed in dramatic ways. Twenty years ago, many local governments received $0.75 in federal revenue for every $0.25 that it spent on wastewater treatment facilities. Those days are gone. Yet the federal government still requires local governments to make expensive wastewater system improvements while no longer providing significant levels of support for expanding facilities to accommodate future new growth. As a result, local government has been reluctant to raise taxes as needed to support new growth and development. This chapter traces the evolution of development exaction policy leading to increasingly widespread use of system development charges.

EVOLUTION OF EXACTION POLICY

Early land use controls focused primarily on subdivision design issues. Alignment of streets, often in accordance with a master street plan, was a key issue. Gradually, local governments developed higher expectations for developers. Many began to require more than merely the construction of streets. They began to require developers to pave streets, install curbs and gutters, and sometimes build sidewalks. As the use of public sewer and water systems spread, local governments began to require that developers install lines under the streets of the new subdivision. There was little litigation over these early exaction practices and little reason for dispute. Clearly, if communities did not require developers to make minimal improvements, they would be faced with paving rutted roads or retrofitting neighborhoods with sewer lines. Paying for such improvements would involve special assessments or general taxes, neither of which was an attractive alternative to local officials.

Subdividers also may have seen increased value accruing to their lots from the installation of such public improvements. A developer lacking such basic facilities in a new subdivision would, after all, have a less marketable project in most urbanized areas. Thus, the direct benefits of the improvements may have been as influential as the obvious need for the improvements in limiting the number of court challenges to early exaction requirements. All of these early exactions involved the dedication of land and the installation of improvements *within* subdivisions. As the regulatory system evolved, however, communities began to go farther.

Improvements Required Within the Boundaries of Development

Development exactions are a product of the evolution in public policy regarding land use and the provision of public facilities. Before the 1920s, most growing communities had no effective land use controls. It was not uncommon to find speculators, for example, subdividing vast tracts of land considerable distances from cities in anticipation that purchasers and home builders would eventually receive city services (Nelson, 1988a). There were no land use regulations controlling the location, timing, or dimensions of those developments, nor were public facility extension policies linked to land use regulation.

The U.S. Department of Commerce's model planning and zoning enabling acts, written in the late 1920s, are the genesis of modern land use regulation. Today, a person can travel to many states and find commonalities in land use regulation that are rooted in the model acts. One important result of the model acts was regulations requiring developers to provide facilities necessary to accommodate development within the boundaries of that development. These are called "on-site" improvements. Prior to the model acts, developers often demanded and received from cities street, water, sewer, and drainage facilities to each part of their development. The model acts gave public officials the policy rationale for requiring developers to internalize that cost (Nelson, 1988b). These early forms of exaction recognized that to protect the public health, safety, and general welfare of the community, developers themselves must be held responsible for the installation of improvements within the boundaries of their development. Such exactions thereby have a *regulatory* role because their objective is the protection of the public. Requiring developers to provide adequate facilities has become commonplace. Modern subdivision regulations require subdividers to provide a number of public facilities as a condition of development approval.

Mandatory Land Dedication and In-Lieu Fees

The four decades following the 1920s saw public officials wrestle with providing facilities outside the boundaries of the development. Many officials realized that fiscal resources could not satisfy the demand created by new development for new parks and schools. In response, many local governments now require developers of residential subdivisions to dedicate land for park and school use (Hagman and Juergensmeyer, 1986). This is usually facilitated by state enabling legislation.

Sometimes, land dedicated by development was in the wrong place, too small, or not otherwise appropriate to satisfy the purpose of the exaction to provide land for parks and schools. Payments of money in lieu of dedication came into use. The amount of money given in lieu of dedication is normally calculated as equal to the value of the land that would have been dedicated (Hagman and Juergensmeyer, 1986).

By the 1940s, local government's power to demand land or money for facilities located off-site was firmly established. But mandatory dedication and in lieu laws did not necessarily enable modern SDCs. This is because in lieu fees are related to mandatory land dedication, but there is usually no mandatory land dedication for water, sewer, drainage, roads, and many other facilities

(Juergensmeyer, 1988). The need for these facilities and services would have to be satisfied on a different, but related, basis.

Extension to Off-Site Exactions

Two forces have influenced the evolution of exaction policy: reconsideration of the growth ethic and local government fiscal stress.

Until the 1960s and 1970s, most communities believed that growth and new development were fundamentally good because they brought an improved tax base that could be used to build better facilities that all community residents enjoyed. In effect, growth meant improving services at declining average cost to taxpayers. Challenges to the growth ethic arose in the 1960s, however, as residents of desirable, rapidly growing communities discovered that unbridled growth caused pollution, congestion of streets, overuse of other facilities, and a general lowering of the quality of life (Reiley, 1974). Cost-revenue studies began showing that, in many cases, new development placed incremental demands for community facilities that increased average tax burdens for existing taxpayers (see Burchell and Listokin, 1978). Many citizens in certain parts of the country, and especially those in rapidly growing areas, concluded that growth was inimical to their reasons for choosing to live where they were (see Scott, Brower, and Miner, 1975). A new land use regulation ethic emerged calling for new development to internalize all the costs it imposes on existing residents (Bosselman, Callies, et al., 1972).

Regulations based on this logic have been found to be within the power of local jurisdictions provided that there is a clear public purpose and that the regulations are reasonable (see, e.g., Callies and Freilich, 1986). Some development proposals may be denied if the facilities needed to support it are lacking [see *Golden v. Planning Board of Town of Ramapo,* 30 N.Y. 2d 359, 285 N.E. 2d 291, 334 N.Y.S. 2d 138 (1972)]. Development is acceptable only when it provides needed facilities. Thus, growth is acceptable in many communities only when it pays its own way. Using this logic, many local governments now hold development at least partially responsible for either constructing or paying for off-site improvements such as roads, parks, fire and police stations, and water-related system expansions.

However, since the late 1970s taxpayers have called for substantial changes in the manner in which local government receives tax revenue. This was manifested in electorates imposing severe restrictions in the taxation of real property on local government. Propositions 13 in California and $2^{1}/_{2}$ in Massachusetts are only two of the more visible outcomes of this revolt (Chapman, 1981). During the 1980s, virtually every state enacted some form of property tax limit that effectively reduced local government's ability to expand infrastructure from taxes.

These tax limit efforts often ignored three factors: (1) the effects of inflation on the purchasing power of tax collections; (2) rising water and wastewater treatment standards; and (3) rising expectations among citizens for higher-quality services. The net result of all these factors is that local governments have been forced to consider all possible revenue-enhancing sources, such as new or higher user fees, privatization of some services, negotiated exactions of new development requiring planning approval, and SDCs.

Refinement of Concepts

Along with the evolution of exaction policy has come an evolution of terms. Recall that exactions have evolved from an "on-site" to an "off-site" orientation. Until recent years, exactions were classified as either on-site or off-site, with most of the policy and legal disputes centering on the extent to which developers should be held accountable for providing or paying for off-site improvements. But this is an inadequate distinction. Developments that generate considerable volumes of traffic may need to have acceleration and deceleration lanes, and traffic signals installed to provide safe ingress and ingress to the development. Yet, these improvements are clearly "off-site". Alternatively, oversized water mains installed by a developer "on-site" may provide greater benefit to future development off-site. A new view is emerging in which improvements are classified as either "system" or "project" improvements.

Generally speaking, "project improvements" are site improvements and facilities that are planned and designed to provide service for a particular development project and that are necessary for the use and convenience for the occupants or users of the project and are not "system improvements". The character of the improvement controls a determination of whether an improvement is a project improvement or a system improvement, and the physical location of the improvement on-site or off-site is not by itself determinative of whether an improvement is a project improvement or a system improvement. If an improvement or facility provides or will provide more than incidental service or facilities capacity to persons other than users or occupants of a particular project, the improvement or facility should be considered a system improvement and should not be considered a project improvement. "System improvements" are capital improvements to the system that are designed to provide service to the community at large, in contrast to project improvements.

Separating the two is not always clear. For example, installation of a main along a highway arterial adjacent to a development project may benefit a local government's entire water system inasmuch as this would clearly fit the definition of a system improvement and it could be eligible for impact fee financing, at least in part. But are right-of-way dedications for water mains that improve service to the development considered project or system improvements? (While a share of the mains improve the distribution network, right-of-ways or easements serve abutting development and are considered project improvements.)

LOCAL GOVERNMENT RATIONALE FOR SYSTEM DEVELOPMENT CHARGES

Consider the taxpayer revolt mentioned earlier. Between 1972 and 1990, there was a movement away from property taxes — caused in large part by voter resistance to the size and regressive nature of the tax. Property tax revenue fell from 36 to 26% of total local revenue. However, there was a 22% increase in debt and increases of 28% in general revenue, 59% in taxes not on property, and 87% in other charges on a per capita basis (Statistical Abstract, 1974 and 1993). The figures thus portray the effect of fiscal revolt — costs were being shifted away from the community at large to new development and users of services. Yet, the level of local taxation and state and local debt still increased faster than incomes. Total state and local government revenue as a percentage of personal income rose from 18% in 1972 to 22% in 1990.

For their part, local governments rationalize SDCs for as many as five major reasons (Frank and Downing, 1988a):

- To shift fiscal burdens from existing development to new development

- To synchronize the construction of new or expanded facility capacity with the arrival of new development
- To subject new development decisions to pricing discipline.
- To respond to locally vocal anti-tax sentiments

The significance of these motivations are now discussed.

Shift Fiscal Burdens

Existing development enjoys facilities that were greatly subsidized by federal, state, and communitywide resources. Nowadays, those subsidies have been reduced or canceled. Federal aid for water and sewer systems used to pay for up to (and sometimes more than) 75% of the cost of those facilities. The community needed only to raise the other 25% or so. When general obligation or revenue bonds were used to finance the community share, the tax burden falling on all residents was very small; but when federal support for these systems was eliminated in the 1980s to reduce the federal debt, communities had to find other means to finance facilities. The trouble was that voters in many of those communities refused to raise their taxes or rates in order to pay for new facilities serving other people.

Also, communities have historically underpriced the use of existing facilities. Prudent facility pricing policies would raise enough revenue to pay for necessary rehabilitation and replacement of existing capital stock. Largely to provide politically popular, low user prices (in the form of direct user fees or indirect taxes), local public officials kept the price of those services artificially low. The result is that, today, communities face a mounting, unfunded capital improvement maintenance and replacement debt that is quite large, by some estimates running over $1 trillion across the nation. Faced with this prospect, occupants of existing development do not wish to have to pay for rehabilitation or replacement of existing facilities, and certainly not for new facilities that over time will also require rehabilitation or replacement. The attitude is that, since existing facilities were paid for by existing development, communities strapped with maintaining or replacing those facilities must require new development to install, maintain, and replace its own facilities.

The political reality is that no one wants to pay the bill for either new facilities or rehabilitation of existing facilities.

Synchronize Facility Expansion and New Development

Leap frog urban sprawl, often considered the most costly development pattern, is caused in part by the extension of facilities into undeveloped areas, allowing development to skip over land closer to existing development. Extension of those facilities in advance of development is also costly if existing development does not immediately use it. Thus, to keep facility costs down, communities now use facility extension policy to control the rate, timing, and location of growth.

Force Pricing Discipline on New Development

Local public officials are coming to accept that underpricing of facilities leads to their inefficient use. Development is less intense, more spread out, and more wasteful of facilities when it does not have to pay the full cost of the facilities to which it connects and uses. Officials now understand that it is more expensive to serve development in some places, less expensive in others, and some kinds of development are simply more expensive to serve than others. With typical user charges and flat-rate connection fees, development in areas that are less costly to serve actually subsidize development in areas that are more expensive to serve. The result is that development in less-costly areas is penalized, while development in more costly areas is rewarded. Maldistribution of development occurs and the total costs of providing facilities rises.

By forcing new development to pay for its full share of the cost of providing new facilities, local officials use market principles to determine when new development is feasible. That is, when facility prices reflect true costs, development that cannot afford to pay those costs will not happen, while development that can afford to pay those costs will occur.

Use Symbolism to Respond to Anti-Tax Sentiments

Local public officials may use SDCs to mollify locally vocal anti-growth or anti-tax interest groups. Sometimes, officials will adopt relatively low SDCs, amounting to a small fraction of the true cost, in order to demonstrate symbolic support of interest group positions. For example, in at least one California community, local interest groups asserted that all new development should be held fully accountable for facility costs, but local public officials adopted some SDCs that reflected only a quarter of those costs. In some other California communities, however, local public officials are elected and re-elected by demonstrating how high SDCs are.

SUMMARY

SDCs have evolved from the combination of local government response to changing needs and political pressures, and judicial evolution of the legal defense of local government actions. Chapters 2 and 3 further develop the judicial, policy, and economic foundations of SDCs.

2 Legal Considerations of System Development Charges

INTRODUCTION

This chapter summarizes the legal considerations involved in applying the *rational nexus* criteria to system development charges (SDCs). It begins with a review of the evolution of impact fee law resulting in the *rational nexus* approach. The chapter then reviews the critical elements of *rational nexus* criteria as generally applied to water and wastewater SDCs.

JUDICIAL BACKGROUND

SDCs are merely another name ascribed to a group of non-tax revenues generally called "development impact fees". Impact fees must pass federal constitutional tests and nexus tests. Although nexus tests are intimately entwined with constitutional considerations, the nexus concept is dealt with separately from the constitutional discussion.

Constitutional Tests

As a class of exactions, impact fees must meet three constitutional tests: procedural due process, equal protection, and taking of property without just compensation.

Due Process

The Fifth Amendment to the U.S. Constitution protects against the loss of life, liberty, or property without due process of law. This has two features: authority and nexus. For water and wastewater SDCs, authority is satisfied in up to four ways:

11

- Explicit development impact fee enabling legislation, wherein local governments are given the clear grant of power to assess, collect, and spend impact fees for a prescribed set of facilities, including water and wastewater systems.
- Home rule powers, wherein states grant local governments broad authority to set fees and charges for the operation of local government facilities.
- The police power, wherein local government is obliged to provide for the health, safety, and general welfare of the public through the regulation of development that may coincidentally require the payment of a fee that is used to provide necessary public facilities.
- Utility rate setting authority as a subset of powers conferred on local water and wastewater systems by state legislatures. Since every legislature has conferred such authority on local water and wastewater systems, it is arguable that all water and wastewater systems have the implied authority to set SDCs. However, states also tend to create different classes of water and wastewater systems, and some may reserve particular rights.

A related issue concerns general taxation. While general authority to impose, collect, and spend SDCs may be available, the manner of assessment, collection, and expenditure must clearly qualify the payment as a fee and not a tax. For example, the City of Columbia, Tennessee, had its water and wastewater fee deemed an unauthorized tax because there was no analysis justifying the fee amount.

To avoid the taxation problem and successfully use explicit or implied authority, it is important to characterize SDCs as a regulation and not a revenue-raising device. The regulatory defense is met when demonstrating that new development creates the need for new facilities; but without payment of fees in exchange for benefits provided by such facilities, the revenues are unavailable to provide facilities and therefore the community is unable to accommodate new development. The nature of this relationship is discussed later in the *nexus* section.

Equal Protection

The 14th Amendment to the Constitution requires that laws treat similarly situated persons equally. In effect, SDCs must be applied to all parties on the same basis, (including federal and state government agencies despite their immunities from paying local taxes and fees). There are two basic elements of equal protection:

- The SDC must not arbitrarily apply to some classes of development, but not to others. All new development that imposes an impact on water and wastewater systems must be assessed the same kind of fee, although fees can vary by the magnitude of impacts. This does not prevent local government, however, from exempting certain kinds of development from payment of fees if there is a clear public purpose for the exemption. For example, low-income housing may be exempted from payment of the fees, but care must be taken to make up the lost revenues from sources not principally related to new development.
- The fee must be rationally related to the public purpose; that is, there must be a rational relationship between the need for new facilities required to accommodate new development and the fees new development pays to finance those facilities. More on this later in the *nexus* section.

Taking

The third significant constitutional test to be met is the taking clause of the 5th Amendment, which is made applicable to states by the due process clause of the 14th Amendment, whereby "private property shall (not) be taken for public use, without just compensation". A SDC charge may be alleged to be a taking if the fee is so onerous as to prevent economical development of the property, or if the fee is considered a form of exaction that itself is not related to a legitimate public purpose. Usually, taking only occurs when government physically invades property such as in the case of requiring an easement across property where the easement does not provide any benefit to the grantor and where the easement otherwise serves no legitimate public purpose [*Nollan v. California Coastal Commission,* 107 S. Ct 3141 (1987)], or using regulation to prevent any economical use of the property where the regulation is not clearly related to a public purpose [*Lucas v. South Carolina,* 112 S. Ct 2886 (1992)].

Since SDCs do not normally result in dedications of property or easements, but rather involve payment of funds, they will normally avoid taking challenges. Moreover, where the fees are rationally based on legitimate public purposes and where the fees meet *nexus* tests, taking should be difficult to establish.

Nexus

To meet constitutional tests, system development charges must meet certain *nexus* requirements. Until recently, there have been three competing *nexus* tests:

- Privilege or reasonable relationship test [*Ayres v. City of Los Angeles* 34 Cal.2d 31, 207 P.2d 1. (1949)]
- Specifically and uniquely attributable test [*Pioneer Trust and Savings Bank v. Village of Mt. Prospect*, 166 N.E.2d 799 (Ill. 1961)]
- Rational nexus test [*Jordan v. Village of Menomonee Falls*, 28 Wis.2d 608, 137 N.W.2d 442 (Wis. 1965) and *Contractors and Builders Association of Pinellas County v. City of Dunedin*, 329 So. 2nd 314 (Fla 1976)]

Reasonable Relationship

The reasonable relationship test is based on California exaction practices dating from before 1950. In the late 1940s, Los Angeles required a landowner, who wanted to resubdivide land along busy Sepulveda Boulevard, to dedicate additional right-of-way to permit widening of the boulevard. The owner refused, noting that the land involved was not even permitted access to Sepulveda Boulevard. The owner's argument was that because the land would enjoy no benefit from the street widening, it should not have to contribute to the cost of that activity. The California Supreme Court sided with the City when it held that:

> In a growing metropolitan area, each additional subdivision adds to the traffic burden. It is no defense to the conditions imposed in a subdivision map proceeding that their fulfillment will incidentally also benefit the city as a whole [*Ayres v. City of Los Angeles* 34 Cal.2d 31, 207 P.2d 1, 5. (1949)].

This is the very test that the California Coastal Commission cited when it unsuccessfully defended its easement exaction policies, which were struck down by the U.S. Supreme Court in 1987. (California now enables impact fees pursuant to the *rational nexus* test discussed later.)

Specifically and Uniquely Attributable

As broad as the reasonable relationship test is, the specifically and uniquely attributable test is narrow. In *Pioneer Trust and Savings Bank v. Village of Mt. Prospect* [166 N.E.2d 799 (Ill. 1961)], the Illinois Supreme Court ruled that an exaction would be permissible if authorized and "[i]f the burden cast upon the subdivider is specifically and uniquely attributable to his activity…; if not, it is forbidden and amounts to a confiscation of private property" (at 833-4). In effect, this test suggests that SDCs could only be justified if based on the precise central facility impacts and exact individual lines serving each new development project. In 1993, however, the Illinois Supreme Court defined specifically and uniquely attributable to mean *rational nexus.*

Rational Nexus

The *rational nexus* test has emerged as the mainstream test to be applied to development impact fees and SDCs. The judicial origin of *rational nexus* is found in *Jordan v. Village of Menomonee Falls* [28 Wis.2d 608, 137 N.W.2d 442 (Wis. 1965)], which upheld park dedication requirements where the City demonstrated that new development creates the need for new parks and thus must contribute a proportionate share of the land needed to expand parks to accommodate new development. In a later case dealing with water and wastewater connection fees that included a capital expansion component, the Florida Supreme Court stated in *Contractors and Builders Association of Pinellas County v. City of Dunedin* [329 So. 2nd 314 (Fla 1976)] that such fees were valid when they:

- Do not exceed that which is reasonably required to fund expansion to benefit future connections
- Are needed to finance expansion that accommodates new development
- Are earmarked for expansion

There has been some concern that the U.S. Supreme Court in *Nollan v. California Coastal commission* creates a new kind of test, which it calls *essential nexus.* One commentator observes that, in this respect:

...*Nollan* establishes a federal exaction test that requires a substantial, rationally linked, and, arguably, direct nexus between permit burden and conditions. An indirect nexus, even if reasonably related, will not pass federal constitutional muster. (Rhodes, 1988.)

Another commentator puts this concern into perspective:

The Court's insistence on a substantial nexus probably does not mean ...a return to the "specifically and uniquely attributable" test ...Properly implemented rational nexus-based impact fees should satisfy (essential nexus) requirements, and it is interesting to note that (essential nexus is based on Florida's rational nexus impact fee case law). (Juergensmeyer in Nicholas, Nelson, and Juergensmeyer, 1991).

Indeed, post-*Nollan* case law has generally supported rational nexus-based impact fee systems. Despite the general reluctance of courts to allow one entity of local government to regulate another, the Utah Supreme Court upheld the imposition of a drainage fee on a school district in *Salt Lake County v. Board of Education of Granite School District* [808 P.2d 1056 (Utah 1991)]. In the context of New Jersey's own approach to the "rational nexus" test — which is really a form of specifically and uniquely attributable test — a mid-level appellate court upheld a plat condition requiring that the developer contribute to the construction of a bridge not located within the subdivision [*Squires Gate, Inc., v. County of Monmouth* [588 A.2d 824 (N.J.Super. 1991)]. Similarly, a mid-level court in Washington state upheld a project approval condition that required the developer to contribute to the improvement of an existing road, including the cost of installing a street light [*Southwick, Inc., v. Lacey,* 795 P.2d 712 (Wash. App. 1990)]. In *Lincoln Property N.C., Inc., v. Cucamonga School Dist.* [280 Cal. Rptr. 68 (Cal. App. 4 Dist. 1991)], the California court upheld a school impact fee.

CRITICAL LEGAL ELEMENTS OF SYSTEM DEVELOPMENT CHARGES

System development charges should comply with the *rational nexus* test. This test requires:

- A connection be established between new development and the new or expanded facilities required to accommodate such development. This establishes the rational basis of public policy.

- Identification of the cost of those new or expanded facilities needed to accommodate new development. This establishes the burden to the public of providing new facilities to new development and the rational basis on which to hold new development accountable for such costs. This may be determined using so-called *Banberry* factors [*Banberry Development Company v. South Jordan City* (631 P.2d 899, Utah 1981)].
- Appropriate apportionment of that cost to new development in relation to benefits it reasonably receives. This establishes the nexus between the fees being paid to finance new facilities that accommodate new development and benefit new development receives from such new facilities.

Failure to meet the rational nexus test may result in a court declaring the fee to be an unauthorized tax or an unallowable exercise of local government power.

Rationally Based Public Policy

The need for new facilities to accommodate new development is best established through comprehensive land use and facility plans. Generally speaking, these plans must include:

- An assessment of existing conditions, including existing facility capacities and weaknesses; weaknesses should include a determination of the extent to which existing facilities are able to accommodate existing demands.
- A projection of future facility needs.
- A determination of the physical and financial dimensions of expanding existing facilities or building new facilities to accommodate new development.
- A land use and capital financing plan to expand or build facilities to accommodate needs.

These planning studies should firmly establish the connection or nexus between new development and the new or expanded facilities required to accommodate it. When this is done, the rational basis for holding new development accountable for some of the costs associated with expanding facilities is established.

To further improve rationality, levels-of-service standards should be adopted and inserted into the comprehensive land use planning and capital improvements program. Levels-of-service standards can be based on domestic water and wastewater flow or treatment impact measured on a per acre, per resident, per equivalent residential unit, per land use unit, per fixture unit, per square foot of different land use types, or other objective basis.

It may be the case that after applying the adopted level-of-service standard to existing development patterns, a deficiency is found. Suppose a local government: adopts a level-of-service standard of 400 gallons per average day for an equivalent residential unit (ERU); has a water treatment plant permitted to operate at 4 million gallons per day (MGD) average daily production; and serves 15,000 ERUs needing 6 MGD capacity. The plant is clearly not producing at local demand levels, at least considering the adopted level of service. The local government has a deficiency of 2 MGD. This deficiency must be remedied by existing users and not by new development.

There may be pitfalls in deciding appropriate facility standards. If they are set higher than current conditions, the community might face considerable deficiencies that courts would allow to be made up only by existing development. If they are set at current conditions and if current conditions are truly unacceptable, then the community must face up to its responsibility to provide adequate facilities to existing development. New development thus cannot be made accountable for providing higher levels of facilities than existing development has been willing to pay for.

Banberry Factors

The *Banberry* Court requires consideration of seven factors to determine the proportionate share of costs to be borne by new development:

- The cost of existing facilities.
- The means by which existing facilities have been financed.
- The extent to which new development has already contributed to the cost of providing existing excess capacity.
- The extent to which existing development will, in the future, contribute to the cost of providing existing facilities used communitywide or nonoccupants of new development.

- The extent to which new development should receive credit for providing at its cost facilities the community has provided in the past without charge to other development in the service area.
- Extraordinary costs incurred in serving new development.
- The time-price differential inherent in fair comparisons of amounts of money paid at different times.

The Cost of Existing Facilities

Determining the value of the current system is important, especially when a local government wishes to recoup from new development some of the investment made by current ratepayers to create excess capacity benefiting new development. This determination can include detailed engineering estimates of the cost to replace the current system, insurance values, or inventorying existing facilities, including their original costs and then adjusting costs based on current dollar factors considering depreciation, further adjusted to reflect the financing costs absorbed by existing development to make capacity available to new development.

How Existing Facilities Were Financed

Existing development must be sheltered from the cost of new facilities necessitated by new development. Conversely, new development must be sheltered from the cost of paying for facilities that will be paid in part by other sources of revenue. The financing of existing facilities must be determined. This can be ascertained from budget records. Typical financing arrangements will include user charges, property taxes, other taxes, revenue bonds (retired by user fees or property taxes), and intergovernmental transfers (from state or federal agencies). Those same financing means may be available to support improvements. The extent to which that is possible must be determined. For example, perhaps 10% of the cost of constructing the water works has come from property tax payments over the past 5 years. This needs to be considered in SDC calculations. Failure to determine how existing facilities were financed and the extent to which such financing will contribute to the cost of new facilities may jeopardize legal defense of impact fee programs. Moreover, grossly underestimating the magnitude of these funding sources through highly conservative assumptions may also fail court review.

How Much New Development Has Already Paid

Owners or developers of raw land have not paid user fees, sales and excise taxes, fuels or other taxes, or fees on that land. They do pay property taxes, however. If property taxes have been used to pay, even in part, for the construction of existing facilities, new development must be credited for the value of those past payments because the undeveloped land on which new development is sited has already contributed to the cost of constructing new water facilities. The present value of those past contributions must be factored into any SDCs. This will usually mean lower fees.

How Much New Development Will Pay in the Future

Future payments made by new development to finance facilities must also be accounted for. Usually, this will be the calculation of the present value of the stream of future payments made by new development to retire debt or otherwise finance facilities. That value will be applied against the SDC.

Credit for Facilities Installed by New Development

Many communities require developers to install facilities or dedicate land in addition to paying SDCs. There is some debate on the extent to which new development should be credited for these facilities. Conceptually, new development may be required to dedicate land for water line right-of-way, install water lines, or extend utilities some distance from the development site. Usually, the value of these other facilities installed by new development that benefit more than the tenants of new development are credited against like-system development charges. For example, Atlanta, Georgia, credits right-of-way dedications against impact fees at 115% of the taxable value of the property being dedicated.

Extraordinary Costs

Development in some parts of a community may be considerably more expensive than development in other parts. Substantially higher-cost areas should be assessed higher fees than lower-cost areas. This may be done through service area assessments discussed in later chapters.

*The Time-Price Differential Inherent in Fair Comparisons of
Amounts Paid at Different Times*

Simply put, the present value of the contributions already made or yet
to be made by new development must be calculated and credited
against SDCs. How this is done is a major element of this book, as
will be seen later.

Benefit

To complete the *rational nexus* test, new development must be shown
to benefit from the fees it pays. This involves two considerations:
assuring substantial benefit and certainty of benefits (see Nicholas
and Nelson, 1988).

Substantial Benefit

The central consideration here is whether new development reasonably benefits from the fees it pays. For water and wastewater systems, this is usually easily demonstrated because the benefit is a
physical connection to the system. For other kinds of impact fees
such as roads, parks, and public safety, the issue is more complex,
especially if such facilities paid for in part from impact fees are
located very far away from contributing development.

Certainty of Benefits

There must be certainty that contributing development will reasonably (although not exclusively) use a facility it helps pay for through
SDCs. This is probably more relevant in situations where fees are
used to build roads or parks. Palm Beach County, Florida, expends
road impact fees within 6 miles of contributing development.

   In the case of water, wastewater, and stormwater facilities, several considerations are at work. First, the physics of water transmission result in reduced system performance if part of the system fails,
even if that part is located many miles away from new development.
Owing to the systemic nature of these systems, it is not uncommon
to view the entire system as benefiting new development even though
SDCs paid by new development may be expended many miles away
from such development.

Moreover, because of scale economies and the lumpy nature of building and expanding these facilities, large amounts of debt may need to be incurred to build and expand them. Debt service covenants normally bind all ratepayers within the same sponsoring jurisdiction. If a county has a dozen treatment plants and the county sells revenue bonds to build a treatment plant and distribution network in just one area, all ratepayers are made responsible for debt service incurred on behalf of that area. Chances are that any given ratepayer is retiring bonds issued to build or expand facilities in many areas. While not impossible, it may be very impractical to set different rates for different areas corresponding to the level of debt service applicable to any given area. Rates for some areas may become so high as to displace development to other areas where there may be insufficient capacity to accommodate it. The entire jurisdiction benefits by distributing new development in a manner that allows for efficient use and economical financing of facilities in any one area. It thus may not be inappropriate in these situations to have one large "service" area for water, wastewater, and stormwater systems. The ultimate conclusion is that when new development pays SDCs and is connected to the system, it receives the benefits of service.

Timing of improvements is also important in showing certainty of benefits. Courts will look to see if SDCs are expended in a reasonable period of time. Reasonableness can be vague. In San Diego, a court-approved facilities benefit assessment program results in collecting fees for facilities that will not be constructed for up to 15 years. Normally, courts will look to see if impact fees will be expended on facilities scheduled for construction within a 5- or 6-year capital improvement program cycle. Sometimes, however, SDCs are not spent in a timely fashion. Therefore, such fee programs usually contain a refund clause.

SUMMARY

Rational nexus criteria assure that new development will not be paying twice for the very facilities financed first by the SDCs it pays, and again by taxes or rates it also pays; but avoiding double payment is only one concern. The other is to assure that new development will reasonably benefit from the fees it pays. This is demonstrated when local government demonstrates a reasonable connection between the fees and the benefits received. In summation, a *rational nexus*-based system development charge must:

- Be rationally based on public policy that demonstrates a nexus between new development and the need to expand or build facilities to accommodate it.
- Not exceed the new development's proportional share of the cost of facilities needed to serve that development, after crediting it for other contributions that it has already made or will make toward that cost.
- Not be arbitrary or discriminatory in its application to individuals or customer classes.

3   Applying *Rational Nexus* Principles to the Calculation of System Development Charges

INTRODUCTION

System development charges (SDCs) should meet *rational nexus* criteria to assure maximum reasonable acceptance by the development community, local government elected and administrative officials, and courts. At the heart of the *rational nexus* test is the concept of "proportionate share", which can be defined as *that portion of the cost of existing and future system improvements that is reasonably related to the demands of new development.* The following general requirements work to ensure reasonable calculation of proportionate share SDCs.

- *Service areas:* system development charges must be calculated based on service areas. SDCs must be expended on the type of facility for which the fee was calculated, assessed, and collected within the service area in which it was collected.
- *Levels of service:* calculation of SDCs must be based on levels of service that are adopted in the local government's comprehensive plan and applicable to existing as well as new development. The level of service should be based on sound planning and be applied uniformly throughout the service area.
- *Capital cost:* the capital cost of constructing new capacity to serve new development is calculated on a "per unit" basis. For water and wastewater, the unit is typically a gallon of treatment. For stormwater, several units may be used, in-

cluding per capita, per square foot of development, or per square foot of impervious surface. This is the average cost of existing facilities and the marginal cost of planned facilities. Nonlocal funding sources must be excluded from these calculations or considered as revenue credits.

- *Demand schedule:* for each land use or, in the case of water and wastewater, each meter size, the demand generated per unit of development must be estimated and then multiplied by capital cost estimates to determine the total cost per unit of development.
- *Revenue credits:* calculation of SDCs must consider the extent to which past and future payments made by new development have been and will be used to finance the same facilities for which new development is assessed SDCs. Normally, existing ratepayers are the only group paying debt service for existing facilities. New development pays debt service only after being connected to the system so that the credit issue applies only to future debt service payments. However, in those situations where property taxes were used to build the existing system or used to finance debt service incurred to build part of the existing system, credit for past and future property taxes must be considered.
- *In-kind credits:* system development charges assessed against any particular development must also consider the extent to which that development makes improvements to the existing system in ways that substantially benefit existing and future uses. The value of such in-kind improvements must be considered as a credit against the SDCs otherwise assessed against that particular development project. This is an issue that is addressed only on a case-by-case basis and is best addressed through administrative procedures.

SERVICE AREAS

Service areas are geographic areas used in the implementation of SDCs. A service area can be defined as *a geographic area established by a local government on the basis of sound planning or engineering principles, or both, in which a defined set of facilities provides service*

Table 1 General Service Area Design Considerations

Facility	Considerations
Water	Water system operates as a pressurized, integrated system, with many redundancies for service reliability. The result is that many water systems best function as one, sometimes very large, urban area.
Wastewater and stormwater	Sewer systems are customarily designed to serve drainage basins. Local government agencies that serve large regions may install separate wastewater and stormwater treatment facilities for each major drainage basin. This practice suggests multiple service area design based on drainage basins. However, delineating service areas according to drainage basins may undermine the nature of pledges to secure debt service incurred to build the systems. Moreover, "pumpovers" and other system "interconnects" cloud the delineation of service areas. Thus, the use of a single service area for wastewater facilities is not unusual nor probably improper.

to development in the area. The service area must satisfy the benefit principles of the *rational nexus* test (see Chapter 2).

What are "sound engineering or planning criteria"? Engineering criteria may include natural or environmental boundaries such as aquifer recharge areas, watersheds or flood plains, drainage basins, or the physical limits of collection and distribution networks. Planning considerations might include political divisions or utility service boundaries. Table 1 reviews general service area criteria.

For SDC calculation purposes, service areas function as "assessment" districts and "benefit" districts, even though no formal districts may be created pursuant to relevant state and local laws. As assessment districts, SDCs may be calculated and assessed on the basis of several service areas within the same jurisdiction. Multiple service areas may reflect differences in desired level of service, the cost to construct facilities, or in the demand generated by new development within each service area. Each service area can only have one schedule, however, to remain consistent with *rational nexus* criteria.

The use of service areas as benefit districts is consistent with the *rational nexus* requirement that the SDCs be spent within the service area from which they are collected. This ensures that the improvements constructed with SDC funds provide reasonable benefit to developments paying the SDCs. As benefit districts, service areas must be consistent with *rational nexus* principles. Service areas must be designed to ensure that capital facilities are built within reasonable proximity to the new developments paying SDCs. The actual distance from a development project to a capital improvement serving it is not important as long as a benefit link can be established. Service areas designed pursuant to sound engineering and planning criteria assure that there is a reasonable relationship between the assessment of impact SDCs on new development and the delivery of facilities benefiting new development.

Once service areas are established and used for SDC calculation, assessment, collection, and expenditure, they may not be easy to change. Moreover, service areas may limit the flexibility with which impact SDCs can be spent. A service area might include many proposed new facilities, but not much developable area. In this case, the SDCs would be quite high because a small amount of development would be financing much of the improvements. Alternatively, a service area might include considerable development potential, but no new facilities. SDCs would be quite low. If a service area is too small, there may never be enough money to finance major improvements. On the other hand, if a service area is too large, some improvements may be so far from the contributing development that it is difficult to show reasonable benefit.

LEVELS OF SERVICE

"Level of service" (LOS) is a *measure of the relationship between service capacity and service demand for public facilities in terms of demand-to-capacity ratios or the comfort or convenience of use or both.* If, for example, water rationing is required every summer, one would conclude that the level of service provided by water facilities is low. However, a local government might be willing to tolerate a relatively low level of service if it is less expensive to provide or if it promotes other desirable policy objectives, such as water conservation.

A distinction is usually made between the actual level of service that is measured at any given point in time and the desired level of service. Ideally, the local comprehensive development plan will state the desired level of service as a matter of local government policy. This becomes the LOS for planning purposes. It is crucial to satisfying *rational nexus* criteria. The relationship between the adopted level of service that is used to calculate impact SDCs for new development and the actual level of service existing at the time of impact fee adoption has important implications in the context of any SDC system. These implications are summarized in Table 2.

If a local government adopts an LOS that is higher than the existing level, existing facilities will be found to be deficient when compared to the adopted standard. New developments will pay SDCs based on the adopted LOS, but will be sharing existing facilities that operate at a lower LOS with current users. As the SDCs are collected and expended, facilities will be upgraded and the LOS will improve for all users. However, new developments would not be receiving the higher LOS for which they are being charged, and existing users would be benefiting from the improved LOS paid for by new development.

Such a situation is inconsistent with the *rational nexus* criteria because it charges new development SDCs that are higher than a proportionate share of the benefits received. It would be inconsistent with the requirement that levels of service must be applicable to existing development as well as the new growth and development, and inconsistent with the restriction on the use of impact fee revenues to finance improvements that create additional service available to serve new growth and development. Thus, if a local government decides to adopt a LOS higher than the existing service level, it must find non-fee revenues to raise existing facilities to the adopted service level. Such revenues should be available based on realistic projections to remedy any deficiencies over a reasonable period of time.

At the other extreme, a local government could adopt an LOS that is below that currently provided. Such an approach would mean that existing facilities have excess capacity available to serve new development. This would entitle the local government to charge new development for its proportionate share of the value of the excess capacity. Such SDCs "recoup" past investments made by the local

Table 2 Implications of LOS Standards

| Characteristic | Adopted LOS compared to existing level of service | | |
	If adopted LOS is below actual	If adopted LOS is same as actual	If adopted LOS is higher than actual
Amount of system development charge	Low	Moderate	High
Future level of service	Decline in future service level to all	Maintain future service at current users	Improve future service to all users
Existing deficiencies that must be remedied by existing users	None	None	Must remedy
Excess capacity	Recoupment of previously incurred investment	None	None

government to accommodate new development. In effect, a local government can use recoupment for revenue enhancement; but setting an artificially low LOS for this purpose alone may be short-sighted. While more of the previously incurred costs of existing facilities would be recaptured, SDCs collected for future system expansion would be limited to the costs of providing the lower LOS. In addition, the lower the LOS that is adopted, the lower the annual amount of impact fee revenues received.

The third option is to adopt a LOS that is identical to the existing level of service. In many ways, this is the simplest and most direct approach. It does not create any existing deficiencies nor result in excess capacity. It simply charges new development the cost to maintain the LOS that existed prior to the development. But it does not address broader policy issues related to conservation.

Choosing Between Peak and Average Daily Demand

For water and wastewater facilities, the LOS determination must be made on either peak or average daily consumption. Although all systems are designed for a certain amount of peaking, many systems decide to use the average daily option for LOS determination. Average daily demand as measured over 1 year is a common LOS choice, but in areas with high seasonal use, such as yard irrigation during summer, the peak month or season is often preferred.

Varying Level of Service by Service Areas

It is clearly possible to have different levels of service for different service areas within the same local government jurisdiction. For example, the local government may be willing to tolerate higher levels of water shortages in suburban areas, where shortages are attributable more to yard irrigation than to domestic consumption. If there is a logical reason for providing more intensive services in a particular part of a jurisdiction, or there are constraints that prevent extending capital facilities to certain areas, such reasons for the different levels of service should be presented in the local comprehensive plan.

Interestingly, the possibility of recoupment SDCs as a revenue source may create a situation where a local government may be able to remedy existing deficiencies. For example, if a local government establishes multiple service areas and adopts an LOS at one level, the existing levels of service within the individual service areas would likely be either above or below the jurisdiction-wide average. Thus, some service areas would have deficiencies, while others would have excess capacity. Until the excess capacity is used up, SDC revenues collected in service areas with excess capacity could be used to remedy deficiencies in other service areas. Used in this way, SDCs could help bring about a more uniform LOS to all areas of the jurisdiction.

Level of Assessment

Sometimes, the amount of the SDC may be viewed by policymakers as too high, and there is some pressure to reduce it. This can be done in two ways: lowering the LOS or charging some percentage less than the maximum.

Lowering the LOS is perhaps more difficult in water and wastewater SDC calculations than for any other facility. Lowering the LOS for parks, for example, merely results in more people using fewer parks. However, consumers of water and wastewater facilities have established patterns of use that are usually difficult to alter. Nonetheless, it is possible to adopt policies resulting in reducing the demand for water and wastewater facilities through pricing, rationing, building and system inspection, reduction of infiltration, and other means of conservation.

Probably the most common approach is to calculate the full cost to serve new development at the existing LOS standard, but then charge SDCs at a fixed percentage of the calculated cost. This approach obligates the local government to provide the adopted LOS without the advantage of receiving all the revenues needed to achieve it. Rates usually must go up.

SELECTING THE ECONOMIC BASIS FOR COST CALCULATION

A critical calculation decision involves the selection of the economic base for cost calculation. There are two general approaches, both of which are presented in this book. The first is *vintage capital* calculation and the second is *marginal cost* calculation.

Vintage Capital

Vintage capital cost calculation is used most commonly where SDCs are a form of equity purchase into a system. These are also called "buy in" fees. Conceptually, the new user becomes an investor in the system and the investment fee is the proportionate share of the system equity. The equity value of the system is essentially the current replacement cost less any amounts not locally paid, such as federal grants, and less accrued depreciation. Since it is an obligation of all users, accrued depreciation must be paid from rates or debt. In vintage capital approaches, however, depreciation recovery in the form of rehabilitation is usually financed from capital reserve accounts financed from rates. The vintage capital approach is useful only when the system has been substantially built out, no major capacity or territorial expansions are envisioned, and depreciation is financed substantially from rates. The approach should also consider the financing costs incurred by existing rate payers or tax payers to provide excess capacity available for new development.

Marginal Cost

Marginal cost calculation is used most commonly where SDCs are used to finance capital expansion as well as to recoup investments creating excess capacity for new development. It is based on the full replacement of the system with no adjustment for depreciation, or the cost of expanding the system to serve new development which is consistent with marginal cost theory. The marginal cost approach is appropriate for situations in which capacity and territory expansions are common and where debt is the primary means of financing expansion and rehabilitation. Adjustments for nonlocal contributions to the system are made only if such revenues are expected to help finance new facilities or future rehabilitation. Chapter 6 presents an example of the marginal cost approach.

Vintage Capital and Marginal Cost Hybrid

Both approaches may be used where there is slow to moderate growth, some capacity and territorial expansions mostly accommodating new development, and rates are used to finance rehabilitation. In this situation, the recoupment portion of the SDC is calculated by the vintage capital approach, and the expansion portion is calculated by the marginal cost approach. The hybrid approach is shown in Chapter 7.

PLANNING REQUIREMENTS

The importance of land use planning is reviewed in Chapter 2. The importance of the capital improvements element (CIE) of the plan, and the capital improvements program (CIP) that implements the CIE in calculating proportionate share SDCs, are reviewed here.

The CIE is a highly recommended prerequisite to the imposition of SDCs. For improvements to be financed in part through SDCs, the CIE should offer an adequate policy foundation. At a minimum, the CIE should:

- Establish future service levels for categories of improvements to be financed with SDCs. Service levels must be defined in quantifiable terms so that the local government's progress in attaining its stated service level goals can be measured.

- Delineate service areas so that everyone involved knows exactly where within the jurisdiction the specific capital facilities and service levels will be provided during the planning period.
- Show capital improvement costs and identify prospective funding sources, including SDCs. The CIE must also separate improvements and costs by service area.

A CIE must include clearly defined service areas, LOS standards, and projections of needed facilities including their costs. In essence, the CIE serves to strengthen the relationship between SDCs and public policy by clearly stating those policies and the role of impact SDCs in effecting them.

A local government should organize CIE components to ensure an orderly flow of information, rational analysis, and a clear understanding of the relationship between infrastructure expansion and the overall goals, strategies, and policies established in the comprehensive plan.

When a CIE is being added to an already approved plan, documentation should be included on how data in other plan elements have been reconciled, amended, or updated. The best way to handle inconsistencies with a previously approved plan is to amend other plan elements while adding the CIE to remove or modify any information that has been superseded. However, if this is not practical for some reason, a local government should list by page or section any data or text that is no longer valid, and indicate why. The CIE should also indicate how data from other elements have been used to arrive at the projects in the schedule of improvements or indicate if more recent or specific data have been employed in developing the CIE. If a CIE is developed concurrently with the comprehensive plan, there should be no problem explaining the relationships between the CIE and information presented in other plan elements. Problems may arise in assuring that CIEs added later are coordinated with the rest of the plan.

CIEs are normally prepared following a three-step planning process to be meet *rational nexus* criteria:

- Preparation of an inventory and assessment
- Determination of policies and needs

- Development of an implementation strategy

Each of these steps is reviewed.

Inventory and Assessment

The purpose of the inventory and assessment is to establish baseline information on the facilities being used, the extent to which existing development is adequately served by existing facilities, the current value of existing facilities, and the extent to which existing facilities are capable of accommodating new development.

Inventory

The ultimate purpose of an inventory is to establish the current and replacement values of the existing system. The current value is important because it helps to calculate SDCs for new development based on the unit value of excess capacity. Replacement value is important because it shows the marginal cost of system expansion, assuming expansion costs are the same as replacement costs. The inventory should list all facilities that are in place and used or reserved for use by development. The list may be organized by year of installation, type of facility, or other means. The inventory need not list every nut and bolt; but keeping the ultimate purpose of the inventory in mind, it is important that all assets are at least grouped into rational categories. Chapter 4 illustrates an inventory approach. The inventory should indicate the year of installation, the useful life of the facility in years, the original cost of the facility, and the amount or percent of that cost paid by local government (i.e., not paid by grants or developer exactions). Items that are obsolete and not being used should be removed from the inventory or put into a special section clearly indicating their status. If an item has been modernized or refurbished, the year of such improvement should replace the original year and the useful life period reconsidered.

Assessment

For SDC calculation purposes, the purpose of the assessment is to estimate the excess or unused capacity of existing facilities and arrive at a current value of those facilities. Estimating excess capacity, if any, is largely a function of LOS standards that are considered in the goals and needs section of the CIE. However, by using existing levels of service, one can easily gauge the ability of existing facilities to accommodate new development. To establish the current value of the system as a whole and the value of excess capacity, the replacement cost of each inventory item must be estimated and then summed across all items after accounting for depreciation. In the absence of engineering estimates on the cost to replace inventory items, one can use an index that accounts, at least in part, for the time value of money since the item was originally placed in service. Depreciation is accounted for by considering the year of installation compared to the expected useful life of an item.

There are some interesting wrinkles about the *rational nexus* fee that must be specially addressed. Marginal cost theory would have the unit value of replacement cost as the fee per unit of development impact. *Rational nexus* principles involved when new development is buying into existing excess capacity requires consideration of depreciation because new development cannot be held responsible for more than a proportionate share of system costs or obligations. This essentially requires accounting for depreciation and calculating the average cost of resulting facility value. Another wrinkle is that, to be consistent with the proportionate share criterion, SDCs based on existing facility value must exclude the value of nonlocal government contributions to those facilities. For example, if the federal government awarded a 75% grant for the construction of the local wastewater treatment plant and that plant has an estimated current value of $10 million (original cost adjusted for inflation and financing less depreciation), then the value of the plant for SDC calculation purposes can only be $2.5 million (which would be the current value of the local government's investment in that plant).

Policies and Needs

CIEs should include a section which is devoted to formally establishing policies for service areas and levels of service, and projecting future facility needs.

Service Areas

A fundamental tenet of the *rational nexus* test is that SDCs must be calculated on the basis of service areas. The delineation of service areas helps to ensure that there is a reasonable relationship between the assessment of SDCs on new development and the delivery of facilities that benefit the development. They may cover all or part of a community; they may even span jurisdictional boundaries. Service areas for various categories of services or facilities may be completely different or they may overlap, but they must be based on sound engineering or planning criteria.

A number of factors may influence the delineation of services areas, and some of these were discussed above. Other considerations might include growth management objectives, economic development strategies, or land use patterns established in the local comprehensive development plan. In short, as long as the respective service areas are defined based on sound engineering and planning considerations, they will likely be considered consistent with *rational nexus* criteria.

Plans must include projections of total residential, commercial, and industrial development by service area. Where there is more than one service area for any given facility category, general land use projections should be made for those service areas using any number of small-area projection or adjustment techniques.

Levels of Service

Level of service policies drive the capital facility planning process. To be consistent with *rational nexus* criteria, local governments must clearly define the LOS to be achieved and maintained for affected public facilities by the end of the planning horizon. Establishing appropriate service levels is a policy decision that should be stated in the form of goals or policies within the plan and the CIE.

Projections of Facility Needs

Unless they are of a recoupment nature, SDCs must be used to finance new or expanded facilities based on projected facility needs. *Rational nexus* criteria require that facility needs be clearly identified. Needs assessments should be based on population projections and employment forecasts developed in the plan and tailored to individual facility categories. Such projections should take into account how extending or upgrading services to various areas within a jurisdiction might affect the local economy, as well as the rate, direction, and quality of development. The provision of services to various areas should also be assessed in terms of impacts on natural and historic resources. Since growth often follows, or even hinges upon, the availability of infrastructure and public services, investments in these items will be a powerful force in realizing the community's future vision for development. Projections of facility needs should include:

- Current levels of service for all facilities for which SDCs are to be charged.
- Determination of whether existing service levels are adequate to meet current needs and an identification of major deficiencies or under-utilized existing facilities.
- Description of variations in current service levels throughout the jurisdiction, such as geographic areas that differ in regard to available capacity, distribution systems, or quality of service delivery.
- Identification of any parts of the community where the provision of services is, or will be, limited by engineering, economic, or environmental factors.
- Identification of areas where new infrastructure will be needed to support local government's desired future land use distribution and promote other goals established in the plan.
- Methodologies used in assessing capital facility capacity needs that are consistent with information provided in other plan elements, such as population projections, economic forecasts, established development densities for various housing types, and land use.
- General description of infrastructure needs for the horizon of the comprehensive plan.

Rational nexus criteria do not dictate which data sources a local government must use for planning purposes. However, once these sources are chosen, they should be used as the basis for determining the projected needs listed in the CIE. The CIE should show how the infrastructure needs of the projected new population will be met. If the population and development elements of a local government's plan support a slow growth scenario, while the CIE describes several infrastructure improvement projects aimed at fostering rapid growth, this would constitute an unacceptable internal inconsistency between the CIE and the rest of the plan.

The needs projection should include a description of infrastructure needs for the entire planning horizon of the comprehensive plan. Project costs and growth projections become more uncertain the further into the future they are extended. However, the CIE is necessary to anticipate long-range needs along with short-range priorities. When a local government is building major facilities, it may be most cost effective to size some types of facilities to meet the needs of the projected population of 10 or 15 years into the future. Other facilities may be designed to be developed in phases. Major capital facility needs for the entire planning period should be anticipated, even if they will not be addressed during the 5-year period covered in the schedule of improvements. While it may not be reasonable to define every project required to meet long-range needs, an overview or general indication of major infrastructure investments anticipated should be included in the CIE. For example, if a local government knows that a new east-west highway is needed within 20 years, this knowledge may affect the planning and placement of other infrastructure in the meantime.

Implementation

Calculating, assessing, collecting, and expending system expansion connection SDCs are actually the last things a local government does. For this reason, SDCs are considered *implementation* devices. Even in the absence of state enabling legislation, most state constitutions indirectly enable SDCs because they are not policies themselves but devices used to implement policies derived under home rule, police power, or planning delegation. Unless recoupment-based, SDCs should be calculated based on a capital improvements program and consideration of sources of revenue available to finance the CIP.

Schedule of Improvements

The CIE must consider system improvements for the entire planning horizon of the comprehensive plan, which for many plans will be 20 years. This means that major long-range projects that will be financed with facility SDCs, but that may not be initiated within 5 years of plan adoption, must be identified or described only in general terms. These projects will often be covered in sufficient detail in the CIE as long-term projects.

The specific capital improvement projects and funding sources listed in a schedule of improvements are not set in stone. If a given revenue source does not materialize to complete a new facility, or priorities shift within a service area so that different projects take precedence at a later date, these changes can simply be reflected in the annual update of the schedule of improvements. On the other hand, changes in service area boundaries or modifications of officially adopted service levels are considered substantial policy shifts that would require a plan amendment.

As part of the implementation strategy for the CIE, the typical CIP schedule of improvements only lists projects (including joint or interjurisdictional projects) to be initiated within the first 5 to 10 years after CIE adoption. However the schedule of improvements is formatted, the CIP must identify the service area(s) of each capital improvement project. The schedule of improvements must include (1) a listing, by year, of all impact fee-related capital improvements to be undertaken over the CIP period after adoption of the CIE, and (2) a listing of all capital projects that will be required to upgrade service levels for existing development within each service area. The listing of SDC-related capital improvements to be undertaken over the CIP period should include:

- Assigning each project to specific service area or areas.
- Brief descriptions of each project. This can be as simple as, "Install 12-inch main and related improvements 2500 feet along State Highway 15 in Service Area 2".
- The time frame for implementation, such as anticipated start and completion dates, although single years are usually satisfactory. A breakdown of a project by phases should be included when: (1) the project phases will occur in different years; (2) part of the project will occur outside the short-term planning horizon; and (3) funding sources or responsible administrative entities are separate and distinct for various parts of a project.
- The amount of additional capacity that will be created to serve new development (if any). Since projects undertaken for the purpose of raising current service levels for existing development must be listed in the schedule of improvements, it is possible that none of the additional capacity of a listed project would be intended to serve new development.

The main idea here is to indicate that the projects listed provide sufficient capacity to serve the projected demand in the particular service area. (Details on how this may be done are shown in Chapter 6.)

Local governments are encouraged to ensure that capital improvement projects proposed for each service area are consistent with any policies stated in the plan regarding the distribution of future growth or differential rates of growth between service areas.

Description of Funding Sources

The CIP should be as accurate as possible in estimating project costs and listing funding sources. If project costs are adjusted or financing strategies change, these changes should be addressed in the required yearly updates of scheduled improvements. To provide the most efficient use of public revenues, traditional financing methods will have to be adapted and combined with impact SDCs. Ideally, the combination of funding sources listed for each capital improvement project should only be established after consideration of various alternatives. The description of funding sources in the implementation strategy should represent the optimum combination that will be to the best advantage of the community.

The description of funding sources in the CIP must include the following considerations:

- Reasonable estimates of total project costs for each capital improvement project listed in the schedule of improvements.
- An analysis of the percentage of each project's stated total cost that is directly attributable to adding capacity to serve new development.
- A description of proposed sources of funds other than impact SDCs that are expected to make up the remaining portion of each project's cost.

If special studies are required to identify costs, and such studies cannot be conducted prior to implementation of the CIP, the studies themselves should be listed as work items in the schedule of improvements, along with the years when they will begin and be completed. If specific project costs are unknown, the local government will need to examine cost data for similar recent projects, seek assistance from experts, or request information from other local governments that have installed comparable facilities.

Local governments are, to some extent, free to define "total project cost" as they deem appropriate. However, consistency should be demonstrated. A local government might establish service level goals that involve expenses that cannot be financed with impact SDCs, such as adding specialized equipment, vehicles, or personnel. In such a case, the total cost of reaching stated goals might include major costs that may be integral to expanding services, but not classified as capital improvements or otherwise not eligible for impact fee financing.

The requirement that CIPs include an assessment of the percentage of total cost that is directly attributable to adding capacity to serve new development is intended to establish the portion of the total cost stated that is actually a "system improvement cost". The portion of the total cost designated as a system improvement cost represents the maximum amount eligible for impact fee financing. It is important to note, however, that the CIP does not mean to require local governments to establish the percentage of the total cost that will be paid for by any particular financing source, including impact SDCs. As a matter of policy, some local governments may choose not to collect the entire costs of system improvements through SDCs. Although it will be important for a local government to calculate the portion of the total cost of each project that will be generated through SDCs when developing an SDC schedule, presenting this level of detail on funding sources in the CIE or CIP is optional.

Sometimes, a capital cost is incurred by a local government for a specific development and is really a project improvement. *Rational nexus* criteria prohibit the use of SDCs to finance project improvements. Thus, the funding source for such facilities must include other than SDCs. Where part of the facility is a project improvement but part is a system improvement, SDCs may be shown as a source of revenue for the system improvement. Ideally, the CIP would indicate the percent of each facility improvement that is a system improvement. Similarly, the CIP should indicate the percent of a system improvement that remedies deficiencies and identify other than SDC revenues to finance deficiencies.

Since the *rational nexus* approach allows local governments to recoup the cost of excess capacity in existing systems, the present value of such existing capital improvement should be stated in the CIP as a total project cost. The existing capacity or demand units available to serve new development should be stated in the CIE, rather than (or in addition to) the original capacity of the project, since some of the original capacity will have been absorbed between the time the capital improvement was built and the adoption of the CIE. Instead of stating that additional funding will come from grants, for example, the CIP should include specific information on the source of grant funds. When local government revenues are to be used, the CIP should specify the actual budget source (i.e., general fund, special option sales tax, revenue bonds, private contributions, or other general categories of financing).

Capital improvement projects required to upgrade service levels for existing development should be included and identified in the CIP to demonstrate that a local government is being consistent with the *rational nexus* approach. In addition to listing projects required to meet service level goals for existing development in the schedule of improvements, specific funding sources should be identified.

Recoupment

The *rational nexus* approach allows local governments with excess capacity to charge SDCs to recover the cost of existing infrastructure that was built before the SDC ordinance was adopted. This is called "recoupment". SDCs based on the principle of recoupment require careful analysis of how and when each applicable capital project was originally financed because, like all fee calculation methodologies, recoupment SDCs must avoid double-charging and include credits that reflect the time value of money.

A community's plan should clearly indicate whether SDC revenue will be used to recover the costs of existing capital facilities with excess capacity. Local governments that plan to recover the cost of facilities or infrastructure already in place should indicate their intention to do so in the capital improvements element of the plan. Communities that use recoupment must also be prepared to document how much service capacity exists at the time of plan adoption. If a local government wishes to assess recoupment SDCs, it must establish a point-in-time estimate of the excess capacity remaining in capital facilities.

Prologue to Analysis

These many considerations are demonstrated in Chapter 6 for wastewater and water SDCs based on vintage capital valuation, in Chapter 7 for wastewater and water system development charges based on marginal cost valuation, and in Chapter 8 for stormwater system development charges based on a hybrid of vintage capital and marginal cost valuations.

4 Administrative Considerations of System Development Charges

INTRODUCTION

In addition to the general planning and methodological requirements discussed earlier, SDCs are subject to many procedural requirements to assure consistency with the *rational nexus* criteria. These requirements are reviewed here in terms of procedural requirements, issues that must be considered in SDC (system development charge) ordinances, and the management structure.

PROCEDURAL REQUIREMENTS

While any given local government has wide latitude in determining the process for developing SDC policy, any such process should include several critical elements to assure reasonable opportunity for public participation and further assure the political legitimacy.

The most common course of action is to form an Advisory Committee to review the need for and craft the structure of SDCs. The committee is usually composed of a spectrum of community interests such as developers, neighborhood representatives, people with technical backgrounds (such as in finance, engineering, and planning), and other broadly recognized interests. A formal, visible, representative process of deliberation is highly recommended. The concern is that failure to pursue a formal, visible, and fair process would invite avoidable community divisions. The Advisory Committee provides general direction and legitimacy throughout the process of SDC policymaking.

Although the involvement of a large committee is not normally efficient for many processes, a reasonably large committee is recommended. For example, the City of Loveland, Colorado, crafted a nationally recognized impact fee program covering all public facilities with an 18-member committee (divided roughly between development interests and neighborhood interests). Many states require committees of certain sizes and compositions. The state of Georgia, for example, requires a committee of not more than 10 members, a minimum of 40% of whom must be from the development community. A reasonably large committee representing the major interests of the community will assure broad-based community input and will normally result in more successful implementation of the program.

The committee must have some staff support. The staff should include the public works, finance, and planning directors and their respective staffs. Indeed, the department heads should be *ex-officio* members of the Advisory Committee. It is their responsibility to provide information for the committee, help it distill and evaluate the information, and guide the committee through the maze of finance law, theory, and practice.

How should the committee manage itself and what are its principal responsibilities? As a starting point, the Committee should be advised of its public policy purposes as defined by locally elected officials. A work plan and schedule should be developed so that committee members are aware of the time frame and can adequately budget their personal resources to be a productive participant.

An initial task should be an assessment of service costs and revenue trends. Interviews may be held with department officials during regular committee meetings. The review of historical cost and revenue patterns may include analysis of factors that contribute to cost increases such as growth, inflation, federal and state regulation, technological changes, current and desired levels of service (LOS), and the nature of local economies of scale.

Although it may not be appropriate for the Committee to become involved in the details of how facilities are planned, designed, and financed, it must be involved in all the major policy issues such as levels of service, service areas, and the schedule for implementing the fees, including the extent to which the fees should be phased in to soften the impact on the development community. The Committee should also work to assure that the established system is simple, provides adequate funding for necessary capital improvements, and is consistent with community growth management policies.

At critical points throughout the policymaking process, the public must be given the opportunity to review and respond to SDC proposals. This is best done through public hearings before ordinances are adopted. Many local governments require at least two public hearings before SDC ordinances are adopted.

A final consideration pertains to the management of the Committee. Although the Committee must be composed of citizens, it is often useful to appoint one member of the local governing body to chair the committee. In this way, the work of the Committee can be more easily dovetailed with the interests of the governing body, which, after all, has to adopt and implement the fee program. Moreover, making an elected official the Committee chair assures that staff will provide adequate support to the Committee, and reduces the possibility that one interest group may dominate the Committee process. Minutes of the Committee meetings should be kept, but staff should be responsible for taking and distributing the minutes.

IMPLEMENTATION ISSUES

System development charge ordinances must contain several provisions that assure consistency with the *rational nexus* criteria and satisfy certain constitutional considerations.

Timing of Assessment and Collection

One issue concerns where in the development process the SDCs are assessed. It is customary to assess and collect SDCs at the building permit stage because the full impacts of development on those systems can be reasonably determined at that time. Sometimes, assessing and collecting stormwater fees at the site-construction permit stage (before issuance of building permits) is preferred. This is not usual practice, however. To assist developers in preparing their *pro formas,* some local governments will issue a nonbinding estimate of fees, with the final determination subject to plan and building permit review.

Nonetheless, there are three reasons for assessing and collecting SDCs prior to the issuance of either a building permit or certificate of occupancy. One is to prevent projects for which building permits have been issued, but that have not been completed or occupied, from avoiding payment of the fees. The second is to ensure SDC assessment and collection for a change of use that does not require a building permit, but does increase demand on the system. The third is to allow fees to be assessed for shell buildings based on the intended use at the time, with the understanding that the fee will be reviewed at the time an ultimate class of user is found (such as office or industrial).

Appeals and Individual Fee Assessment

SDC ordinances must include an appeals process to address administrative decisions. The appeal may be to the governing body or another body designated in the SDC ordinance. A developer may also pay the fees under protest to obtain development approval, while retaining the right to appeal and the right to any refund for any fees deemed illegally collected. The ordinance may provide an option for resolution of conflicts over the amount of the fees through binding arbitration. In general, there are two kinds of administrative decisions subject to appeal. First, applicants choosing to use the fee schedule may disagree with the administrative classification of the land use applicable to the proposed project. Second, applicants choosing an individual assessment may appeal the SDC resulting from that process. The whole purpose of the appeals and individual fee assessment provision is to ensure that projects sufficiently different from the norm are afforded an opportunity to have fees tailored to their special circumstances.

Administrative and Accounting Requirements

SDCs should not be comingled with general revenue or enterprise funds. *Rational nexus* criteria are best satisfied when fees are spent for the category of system improvement for which they were collected and in the service area in which the project for which the fees were collected is located. Separate accounts should be established and maintained in sufficient detail to identify revenue collections, expenditures, and obligations by specific capital improvement, by service area, and by individual development. Fee accounts need only be ledger arrangements, but some local governments may wish to establish separate bank accounts or trust funds to assure integrity.

Accounts should be established for each service area for each category of facilities. However, where the entire fee is related to recoupment, those fees may pass through the SDC account to any other account as determined appropriate by the local government.

Annual Report

The local government should prepare an annual report showing what fees were received, what interest has been accrued on the fees, the amount of fees that were expended during the year, and the balance of fees in the account at the end of the year.

Refunds

Inasmuch as SDCs cannot be considered a general tax (at least without special state enabling legislation), feepayors must be entitled to a refund of fees if they are denied service. SDCs should also be returned to the feepayor, with interest, if they have not been spent or encumbered within a reasonable period of time after payment. It is customary for ordinances to require expenditure of fees within 6 years of collection in the service area from which they were collected, but periods of 10 or more years are also seen.

Ordinances should specify that fees are deemed expended on a first-in-first-out (FIFO) basis. To determine whether SDCs paid by a particular project have been spent or encumbered within the time frame specified in the ordinance, the FIFO concept is used. Upon determining that a right to a refund exists, the local jurisdiction should provide notice to the fee payor and publish such notice within a certain number of days (usually 30 days) of the expiration of the expenditure limit. An application for a refund should be made within 1 year of the expiration of the time limit or the date of notice publication, whichever is later. Refunds should be made within 60 days of receipt of a valid application and should include a pro rata share of the interest earned on the impact fee account.

Entitlement to refunds may not "run with the land". Refunds may be made to the person or entity who initially paid the fee unless the right to any refund has been transferred to a successor in interest in the property. As a practical matter, communities rarely fail to expend fee revenue in a timely manner if service area definitions are broad enough to allow expenditure on a wide range of projects.

Construction Credits

Credits for in-kind or monetary contributions made by individual developers are often called "construction" or "in-kind" credits to distinguish them from "revenue credits". Revenue credits reflect future payments common to all feepayors, such as property taxes or utility rates, and such credits are reflected in the SDC schedule. In contrast, construction credits are deducted from the fee calculated from the fee schedule on a case-by-case basis.

Construction Credits

The SDC ordinance should allow local governments to enter into private agreements with developers or consortiums of developers to build or provide system improvements eligible for financing with SDCs (even with a value in excess of the fees due on their projects). They may receive credits for the present value of these improvements against SDCs due on the same category of capital improvement. Credits may be transferred to other projects in the same service area. Where these improvements are in excess of credits due against SDCs, the agreement may specify that the developer be reimbursed for extra contributions from the pool of SDCs collected from the same service area. The ordinance should specify that local governments are not responsible for reimbursing a developer for exactions or dedications out of government revenues other than SDCs. This allows new development to benefit from the completion of systems improvements in advance of the additional new growth required to build a large SDC pool. On the other hand, the developer (not the local government) accepts the risk that he/she may not recollect his/her money rapidly if the system improvements do not stimulate the anticipated level of development in service area over the short term.

Credit for Abandoned Projects

An ordinance should provide that if a development permit is abandoned, future development on the same land should receive credit against future impact fees for the present value of impact fees already received from the previous developer. The local government keeps the fees because it has very likely used them to pay for facilities serving the abandoned project.

Exemptions

Some ordinances allow for exemptions from the payment of SDCs for all or part of particular development projects that create extraordinary economic development and employment growth or affordable housing. As with any system in which government collects revenues, various groups sometimes seek exemptions from SDCs. The problem is that if the pool of exemptions becomes too large, the whole purpose of the SDC policy is undermined.

Federal agencies are exempt from payment of SDCs under the separation of powers doctrine, and the local government may not be able to compel payment by state agencies. Despite whether it can actually collect the fee, however, the local government should assess SDCs on all new development in order to be consistent with *rational nexus* criteria.

The policy supporting such exemptions should be contained in the local government's comprehensive plan. In addition, the exempt development's proportionate share of system improvements should be funded through a revenue source other than SDCs. (However, this consideration need not apply to "recoupment" fees.)

Interjurisdictional Agreements

The ordinance should enable the local government to enter into legal agreements with other local governments, service districts, authorities, or with the state to develop joint plans for capital improvements and to jointly collect and expend SDCs. Local governments wishing to enter into intergovernmental agreements to impose, collect, or expend fees should first complete the comprehensive planning requirements implied by the *rational nexus* criteria, and then adopt their own SDC ordinances with fee schedules and interjurisdictional agreement clauses. This ordinance should include a provision that allows local governments "jointly affected" by large or intensive development to collect SDCs to maintain service levels threatened by "spill-over" effects. SDCs may also be collected through interjurisdictional agreement to pay for certain types of systems improvements that traverse jurisdictional boundaries, such as storm water retention reservoirs. In general, an intergovernmental agreement should provide for:

- Joint planning of the capital improvements element
- Joint preparation of the SDC
- Assignment of responsibility for constructing or contributing impact fee revenues for capital improvements
- Administration of the SDC program
- Notification prior to termination or change in the agreement by either party, including provisions for disposition of funds

Management Structure

The SDC ordinance should designate a particular local government office as the fee administrator. In smaller local governments, this can be the chief executive officer. In many larger local governments, the administrator is the planning director. Less typical administrators are building officials and utility directors. The ordinance must afford the administrator discretion in interpreting the ordinance because the ordinance will be unable to anticipate every conceivable situation. The designation of an administrator can shelter elected and certain senior officials from the day-to-day complexities of the SDC program. Nonetheless, appeals of the administrator's determinations should be allowed to be presented to those officials. This meets the due process and administrative appeals requirements of the *rational nexus* approach.

5   Implications of System Development Charge Policy

INTRODUCTION

There are some who argue that SDCs are bad public policy because of several perceived adverse effects. This chapter begins with a review of those concerns; but first, a basic economic lesson.

The market for development is composed of landowners, land developers, builders, and consumers (property owners and tenants). SDCs will be paid by one or all of these participants. The question is, *who?* In the short term, land developers and builders will pay fees out of their profit — and this can be unfair to them. This is because, try as they might, they cannot raise the price of their product. Homebuyers will not pay a higher price reflecting the impact fee if they can buy a comparable house for a lower price in another community.

The trouble is with the timing of the fee. If the fee is imposed before developers have had a chance to account for them, developers will have to pay the fee out of their profit. A common solution by local government is to phase impact fees over a period of time so developers have a reasonable chance to account for them.

After about 2 years, the development industry will have found ways to accommodate fees without raising prices above competitive levels. According to classic economic theory, it will be landowners who will absorb the fees by being forced to lower their price of land. Only if fees are very high will landowners be unable to absorb the entire SDC costs.

There are several myths about SDCs that are inconsistent with basic economic logic. Such myths are usually advanced without economic and analytic foundation. Often, the myth so distorts reality as to confuse the central purpose of SDCs to *improve* the climate for development. These myths fall into categories concerning effects on housing and development prices, equity to low- and moderate-income households, economic development, competition with other suburbs and other major cities, and administration.

MYTH 1: SDCs Will Be Passed on to Homebuyers

Some may argue that SDCs will be passed on to the homebuyer. Many public officials accept this. This is not true. In a recent *Atlanta Journal-Constitution* article, for example, one developer alleged that a $1400 impact fee will result in a $15,000 increase in the price of a house. Yet, if housing prices could rise by $15,000, why wouldn't developers charge the higher prices anyway? If SDCs could be passed along to homebuyers, it means that homebuyers are willing to pay a higher price for the same house. This implies that house prices are lower than people are willing to pay and developers are "leaving money on the table". Of course, developers make sure that they are charging as high as they can for their products.

The real answer is that in the short term, if the fee is not properly phased in to give developers time to respond to the financial effects of the fee, it is developers who will pay the fees. This is one major reason why developers so strongly oppose fees.

If the fees are properly phased, then developers will not be caught in the short term by paying for fees out of their profit. However, landowners will begin to oppose fees once it becomes clear to them that their prices are lower because of the fees.

MYTH 2: SDCs Are Bad for Low- and Moderate-Income Housing

Developers and some community groups may argue that SDCs will be bad for the production of low- and moderate-income housing. This is true if the fees result in reducing land prices to nearly zero. But at the level of fees usually assessed, the fees will have no effect and can actually stimulate production. How can this happen? Another simple economic lesson is needed.

One common way for housing prices to rise is when demand exceeds supply. World Series tickets are so expensive when sold by scalpers because there aren't enough tickets available to meet demand. What causes supply of buildable land to fall below demand?

Indequate infrastructure. What is the purpose of SDCs? To assure provision of adequate infrastructure. In effect, SDCs enable local government to increase the supply of buildable land to more closely match demand. If supply meets demand, prices will not rise.

Thus, there are two dynamics at work to keep low- and moderate-income housing prices competitive. First, SDCs will result in lowering land prices to offset the fee. Second, SDCs will increase the supply of buildable land, thereby also dampening price effects. Indeed, careful economic analysis of SDCs applied to competitive housing markets will likely show improvement in opportunities for low- and moderate-income housing. Nevertheless, communities can waive SDCs on low- and moderate-income housing if paid from other sources.

MYTH 3: SDCs Will Have Border Effects

The argument goes that if a developer is choosing between two parcels of land on which to build — where the first parcel is inside a city where SDCs are charged and the second is just outside where lower or no SDCs are charged — the developer will choose the second parcel.

The trouble is this means that the owner of the first parcel does not make a sale. The landowner must lower the land price to offset the fee in order to make a sale. However, if the landowner does not lower the price, this indicates that the value of future development may be higher on that parcel. Thus, be wary of developers who claim they will choose the second parcel. Chances are they would not have chosen the first parcel anyway. In the meantime, the land market will be holding the first parcel available for higher value development. In effect, what might look like a loss in the short term may be a much higher level of development in the long term.

MYTH 4: SDCs Are Bad for Economic Development

The argument goes that because SDCs raise the price of doing business, they frustrate economic development. However, just the opposite is usually true. First, remember that SDCs will be offset by reduced land prices and by enabling the community to more easily expand the supply of buildable land relative to demand.

Now, consider what economic development *really* looks for: skilled labor, access to markets, and land with adequate infrastructure. Competitiveness for economic development will be stimulated by the new or expanded infrastructure paid in part by SDCs. Besides, local governments retain the option to waive SDCs for specific kinds of economic development, such as development locating in enterprise zones. In the competition for certain kinds of development, it will be able to show developers the dollar value of SDCs waived as a solid demonstration of the local government's commitment to such development.

MYTH 5: SDCs Are Cumbersome and Difficult to Administer

Just the opposite is true. Once in place, SDCs are the most efficient method of exaction. Developers pay fees based on a published fee schedule. Gone is the time-consuming, unpredictable, usually unfair horse trading between developers and local government for system improvements. Instead, all developers are dealt with equally and predictably.

From economic and developer perspectives, SDC programs are more *efficient* than current system improvement exactions, more *predictable* for all developers, and more *fair* because all developers are dealt with equally. Moreover, because of *certainty* in exactions and *economies* in processing approvals that will reduce time involved in securing development approval, the entire development process becomes more *streamlined* and *competitive*.

System Development Charges and Ethics of Public Policy

On other dimensions, SDCs call into question basic public policy ethos, often with sound reasoning. As such, SDCs may be viewed as bad public policy unless they are made to be sensitive to certain policy considerations. Those considerations are reviewed here.

The Inherent Regressivity of System Development Charges

SDCs are fixed charges assessed on new development. Their amount does not vary by value of development or the income of those using occupying new development. SDCs are regressive because they impose a greater financial burden on those with less ability to pay.

However, where such fees are based on fixture units installed in structures, they can become *proportionate* because more expensive homes, for example, will usually have more fixture units than less expensive homes (see Chapter 7).

The Exclusionary Effects of System Development Charges

SDCs can be viewed as an attempt to enhance the community's welfare by attempting to exclude certain types by development and social groups (Frank and Downing, 1988). In this respect, SDCs have an effect similar to exclusionary zoning. It could become easier to demand that new development "pay its own way" without making it seem as if moderate-income families with children are being zoned out (Lee, 1988). However, such exclusionary effects on housing can again be offset by tailoring fee schedules to fixture units.

Excess Benefits Accrue to Existing Residents

It can be argued that SDCs fund improvements that upgrade the entire community, providing new facilities that existing residents use, but for which they have not paid. This would be, strictly speaking, a violation of SDC principles because new development can only be assessed its proportionate or fair share of the cost of new facilities. Nonetheless, the addition of new facilities paid for by new development can generate incidental benefits to the community as a whole, and that can mean the community is upgraded at the expense of new development (Callies and Freilich, 1986). Moreover, infrastructure improvements generate benefits to existing property owners in the form of increased property values because it is primarily infrastructure that creates new urban land value. This increase in value enables communities to finance more improvements with no increase in property tax rates. New residents may pay more than their fair share of the benefits they generate for the rest of the community.

As a result of SDCs, some communities will gain, but at the expense of others. Communities with real estate markets that are modestly insensitive to price barriers to developer entry gain three windfalls. First, the property tax base rises as prices rise to pay the fee. Second, as prices rise, only more affluent occupants move into the community, thereby increasing taxable sales throughout that

community. Third, as affluent occupants displace lower-income occupants, community expenditures for indigent services fall. Other
communities will bear the burden of displaced development and
increased fiscal stress (Huffman, Nelson, Smith, and Stegman, 1988;
Blewett and Nelson, 1986). Yet, in most circumstances, such communities are likely, already exclusionary.

Intergenerational Impacts of System Development Charges

 Snyder and Stegman (1986) suggest intergenerational effects of
system development charges. If SDCs raise the price of new housing in a community, comparable existing housing will also increase,
creating a windfall for owners. If SDCs result in lower house quality
or higher density in order to keep the housing bundle affordable to
otherwise price-sensitive buyers, the price of existing housing at
better quality and lower density will rise, thus creating a windfall
for owners. Only if SDCs do not raise the price of housing will
there be little or no intergenerational effects. Interestingly, it is
possible that public officials may wish to have SDCs result in
higher house prices and windfalls to owners of existing development. The justification is that existing owners are being compen-
sated for either their loss of welfare associated with new development or the past financial burdens they incurred to pay for facilities
that now serve new development (Blewett and Nelson, 1988).
However, when existing residents wish to relocate within the community, they may roll their windfall profits into another home in the
community.

Lack of Political Representation

Newcomers who might object to the fees because they cause higher
prices or rents are not effectively represented in the political process
at the local level (Beatley, 1988). In effect, exactions often represent
efforts by residents and the broader jurisdiction to impose the costs
of public improvements on new development and homeowners who
have had little or no opportunity to participate in the political process
(Butler and Myers, 1984). On the other hand, the development community usually represents newcomers in their effort to avoid paying
the fee.

System Development Charges Represent a Shirking
of Public Responsibility

Most developers feel that SDCs represent a shirking of public responsibility for financing the infrastructure necessary to support new developments (Porter, 1988). The argument is that only through collective action can all of society improve the quality of life. The spirit of collective action, wherein everyone contributes a little but no one absorbs large costs, is undermined by SDC policy.

Since, under the new ethos, new development is charged much more than existing development to locate in the same community, one can only guess whether new residents to the community will be more or less willing than existing residents to impose on themselves higher taxes to pay for community-wide facilities. New residents may protect themselves by arguing that they have already paid more than existing residents for some facilities, and so why should new residents pay for new facilities that benefit existing residents? (Nelson, 1987.)

The argument challenges some of the important purposes of planning. On the one hand, SDCs permit planners to play an important role in establishing a substantial fiscal base for communities without overburdening current taxpayers. However, in fulfilling their important roles as fiscal agents, "system development charge planners" essentially define the public interest to include only the needs of the community for which they currently work. Such planners eschew a role that seeks to address broader social issues, such as housing inequities, and thereby reduces planning to a narrow, technical role (Connerly, 1988).

System Development Charges as a
Win-Win Policy

In practice, SDC policy is favorable to many views for four reasons. First, SDCs are, in fact, more consistent with public finance criteria in practice than most other alternatives to financing new facilities. This is shown by reviewing how SDC policy is applied to be legally defensible. Second, SDC policy generates more benefits to the development process than the status quo or other real alternatives to managing development pressures without resources to manage it. Third, there are favorable implications of SDC policy on the distribution of development within and among communities. Fourth, the relationship between SDC policy and improvements in public policy along ethical dimensions must not be understated.

System Development Charge Policy-Making in Practice Satisfies Public Finance Criteria

It is important to note that the preferred methods of financing public facilities have several important practical and political limitations, despite their theoretical advantages over SDCs.

SDCs can be viewed as a facility pricing instrument that guides development timing and location more efficiently than many alternative policies. If the cost of paying for new facilities is to be borne by new development, one may ask whether SDCs are the most appropriate method. The alternative financing answer is that there are a number of approaches other than SDCs. However, SDCs can be the best approach if one considers that other alternatives suffer from many practical limitations, SDC design incorporates several important elements of theoretically best approaches, and SDCs are administratively efficient.

The choice of the best facility financing method is or should be based on characteristics of marginal cost, price elasticities of demand, scale economies, service areas, exclusivity of use, and internalization of externalities (Lee, 1988; Fisher, 1989). In general, SDCs are most appropriate only when:

- Marginal costs (or long run average costs) are not lumpy, but rather smooth across different levels of production.
- The price elasticity of demand for the facility is high, meaning that people can change their behavior (or volume of facility capacity consumed) without a reduction in welfare relative to suitable substitutes.
- Scale economies are relatively constant (or at least do not take on characteristics of natural monopolies).
- Service areas can be clearly defined.
- Consumption of the facility capacity is restricted to only those who pay for it (there are no free-riders, as in the case of suburbanites using downtown parks).
- The benefits or costs of the facility are taken into full account by users of the facility.

Water, wastewater, and stormwater SDCs substantially meet those public finance criteria, but there remain many additional considerations.

Consider that SDC policy design requires the application of several public finance criteria to be legally defensible. SDC assessments are based on projections of demand by service area, calculation of costs to accommodate demand, and apportionment of those costs to different kinds of development based on the different levels of facility capacity consumed by such development. In these respects, SDCs allow local government to take improved account of: marginal costs (or long-term average costs) inasmuch as fees can vary proportionate to actual costs; price elasticities of demand because fees can alter locational behavior so that high-cost areas are not subsidized in development by the financial contributions of low-cost areas; scale economies because costs will vary by the particularities of the geography being served; and service areas that are designed based upon these principles. Exclusivity of use is achieved in the case of water and wastewater facilities. Finally, because users are physically connected to water, wastewater, and stormwater systems, SDCs substantially internalize externalities in ways probably superior to most alternatives.

Consider also administrative efficiency. One of the major advantages of SDCs over alternatives is the relative ease with which they can be designed, adopted, implemented, and changed. They have the significant advantage of being administratively efficient. Most growing communities, no matter how large or small, can administer SDC programs with a plan, a capital improvements program, a set of forms, an inexpensive microcomputer, and a regular line administrative person. Moreover, SDCs finance the very facilities benefiting development that pay them. Without them, new development may not proceed; or, if development is allowed with commensurate expansion of facility capacity, quality of life may deteriorate. In these ways, SDCs may be the best solution.

Favorable Effects for Developers

Most commentaries on both the positive and negative aspects of SDCs ignore the reality that, to be legally defensible, they must adhere to important public policy principles. By themselves, SDCs may indeed have some adverse effects. Yet, SDC policy is fundamentally rooted in an entire land use planning, facility programming, and fiscal accounting framework. In this context, the decision to embark on SDC, is actually a decision to engage in detailed planning and financing of urban development. Once planning, programming, and financing is accomplished, several important and highly beneficial outcomes emerge.

Payment of SDCs establishes a contract between the developer and local government (Nicholas, Nelson, and Juergensmeyer, 1990). In return for the fee, local government promises to deliver public facilities and services more or less on demand, even if they are not present at the time the fee is paid. This has important implications for accommodating market demand for development. In effect, SDCs are really a positive factor in the production of housing and other kinds of development. To some, this may seem contradictory since SDCs are often popularly viewed as a negative factor of production. However, consider the role of SDCs in the development process *after* they are fully in place. Six beneficial outcomes accrue to developers:

- Communities can only apply SDCs after they have done their land use and facility planning, and programmed the timing and location of facility improvements. Planning and programming creates a climate of certainty for developers. A given parcel of land is known to have developable parameters. Necessary facilities are either in place or will be in place according to a time frame. The SDC is viewed as a direct investment by the developer into needed infrastructure. It is viewed by the community as a commitment to provide that infrastructure because the money has or will be paid.
- The time involved in processing development proposals should be reduced in communities that employ SDCs. Since plans and programs are already in place, only developments requiring greater facility capacity than planned or programmed will be caught in lengthy development review processes. Where SDCs are used, the time involved in project review is usually shortened, although other issues pertaining to design rather than density or intensity may still delay project approval.
- The political attitude toward development may be changed. In communities without SDCs, citizens and taxpayers are rightfully concerned that new development will ultimately

raise taxes. SDCs are a political response to the new ethos that requires development to pay its way. By implementing SDCs, communities greatly diffuse political opposition to development. This can only improve working relationships between developers and the community.

- SDCs usually result in replacing the ad hoc apparatus of negotiated development exactions. This gives developers further certainty, reduces development review time, and does not result in one developer negotiating a better deal than another for similar projects.
- SDCs can improve developers' financing package inasmuch as the uncertainties of adequate facilities or questions of appropriateness of development densities or intensities are substantially, if not totally, eliminated. A financier is probably more willing to approve a development loan package with SDCs than without because the community assessing SDCs has already determined the appropriate use of the property and programmed facilities to serve it. These questions may be unresolved in communities not assessing SDCs.
- SDCs are really nothing more than a one-time facility capacity reservation fee in that, in return for the fees, facilities are delivered. That cannot be said for taxes or even for many other kinds of fees developers still pay. Indeed, the value of the fee is returned perhaps by several times because capital improvements serving land are often the single greatest contributor to land value.

Consider the alternative. In the absence of the new revenue generated by SDCs and considering state and federal cutbacks, the fiscal revolt, declining purchasing power of the tax dollar, rising facility performance standards, and rising demand for higher quality facilities, communities face real dilemmas. Should they stop or slow development; or reprioritize limited facilities budgets, perhaps shifting funds from maintenance and operation to new facilities (as seems common practice); or raise taxes? For all too many communities, the choices fall into development moratoria, or growth controls, or SDCs. Development moratoria are not a permanent solution and will ultimately fail judicial scrutiny. Growth controls will have the effect of reducing the supply of buildable land and result in raising development prices accordingly. Only SDCs, which by definition raise new

revenue to pay for new facilities that accommodate new development, have the potential effect of increasing buildable land supply
and facility capacity commensurate with demand.

Favorable Implications on the Distribution of Development

The practice of SDC policy has favorable effects on the production
of low- and moderate-income housing, distribution of fiscal benefits,
and distribution of regional development. However, these favorable
effects may be undermined by other local government policies that
aim to improve welfare through growth controls or other mecha-
nisms. By themselves, SDCs in these dimensions assumes that SDCs
are fundamentally intended to improve the efficiency of a community in accommodating development while not charging development for more than its proportionate share of facility costs.

Low- and Moderate-Income Housing Effects

Taken on the whole, SDCs and all the planning, programming, and
political certainty they represent are a positive factor in the production
of housing and other kinds of development. Take, for example, the
effect of SDC policy on the provision of low- and moderate-income
housing. SDCs help pay for the very facilities needed to expand such
housing stock. One of the great impediments to providing affordable
housing is infrastructure. Without infrastructure, development is delayed. Delay causes restrictions in the supply of buildable land. This is
turn frustrates demand, thereby raising land prices and, ultimately,
house prices. SDCs actually ameliorate the price effect of infrastructure limitations. Land supply is brought into the market reasonably
contemporaneously with demand because new development supplies
the cash needed to extend infrastructure to new areas for development.
Not only is supply increased to match demand, but questions about the
efficacy of development are removed when SDCs pay for facilities
pursuant to a plan for developing new areas. When new land is made
available for development, questions on which kind of development
are already answered through the planning process.

An example is illustrative. The City of Orlando, Florida, has
approved a several thousand-unit, mixed-use housing development
to be built-out in phases over two decades. It is called Timberleaf.
Inasmuch as all the major questions of scale, timing, facilities, and

density were resolved in one master plan executed in a development agreement between the developer and the city, the developer can receive final plat approval for subdivisions within the project within 30 days. In contrast, the average subdivision in Florida takes nearly 2 years to process (York, 1991). The developer builds houses as the market demands. In the early 1990s, housing prices ranged from the low $50,000s to the middle $70,000s. Not only are the house prices affordable by most standards, the developer also pays $5000+ in SDCs and other impact fees. The combination of certainty in development approval and certainty in infrastructure availability allows the developer to provide lower-cost housing in shorter periods of time than competitors.

Regional Growth Distribution and Fiscal Effects

In terms of regional effects, it may very well be that communities with SDCs continue to grow at historic rates and development per se is not shifted. The reasoning is that because system development charges enable buildable land supply to expand more rapidly than in the absence of SDCs, development is accommodated rather than delayed or shifted. Moreover, where development is shifted, it may be shifted to communities that have excess capacity in existing systems. It can also be argued that SDCs may have favorable fiscal effects on communities that employ them. Overall community fiscal resources are increased because SDCs accommodate development thereby increasing the local property tax base.

There is another consideration, however. In growing metropolitan areas, those areas having certain advantages of centrality or amenities will increase in value. It is unreasonable to assume, for example, that the addition of 1 million people to the western Los Angeles region during the past decade would not increase the price of housing (after adjusting for inflation) in desirable neighborhoods. Indeed, homeowners who sell their homes in this area often cannot afford to buy comparable replacement housing. It is inappropriate to associate SDCs with a shift in housing demand toward more affluent households in these situations. With or without SDCs, markets that are observed to have shifted toward more affluent households would have done so anyway. Indeed, it is possible that were it not for SDCs that help increase the supply of buildable land, the shift toward more affluent households would have occurred sooner or in greater magnitude.

And where do lower- and moderate-income households go in this situation? One need only consider the land rent gradients of traditional urban land economic theory to understand that they are displaced to locations farther away. To the extent that communities farther away anticipate increased demand associated with regional growth and shifting markets, displaced lower- and moderate-income households will be accommodated. Indeed, the greatest danger within growing regions is not that some high-cost communities impose SDCs, but that other communities that are likely opportunities for growth do not impose SDCs and are not equipped to manage growth when it comes.

If communities in rapidly growing regions do not properly anticipate growth, whether they use SDCs or not, fiscal resources will be tested. If growth is properly anticipated, such communities can expect to be better able to match growth pressures with provision of public facilities and services, and, in so doing, improve their fiscal structure. On the other hand, communities that see a shift in household composition toward greater affluence would probably see this shift with or without SDCs. Their fiscal structures would be improved in any event.

Favorable Implications of System Development Charges in Terms of Policy Ethics

SDCs need not be considered "bad" public policy when viewed in the context of the overall planning, programming, and financing rigor that precede SDC policy.

Regressivity

The "old" school of property taxation theory alleged that such taxes were regressive, but the "new" school demonstrates that such taxes are probably proportionate. SDCs can be regressive because they are a fixed cost that does not vary by house price. However, the potential regressivity of SDCs can be addressed in a number of different ways. First, without SDCs and in the absence of alternative sources of infrastructure financing, the resulting effect is that a community moves more rapidly toward affluent household composition and exclusivity inasmuch as land supply is constrained relative to demand. Second, some forms of SDCs indeed vary by the size or type of housing unit. Some SDCs are based on the number of plumbing fixtures. In these cases, SDCs are lower relative to the larger homes purchased by more affluent households. Third, SDCs of the future may be more explicitly based on the size of housing units owing to recent evidence suggesting that larger housing units demand more facility capacity for many facilities than smaller units (Nicholas, 1991). Fourth, communities have the clear option to waive SDCs for certain kinds of housing provided the waiver meets other public policy objectives and makes up the fees from other sources of revenue.

Exclusivity

SDCs do not by themselves make a community exclusive. This is the result of other policies or market dynamics. Exclusive policies include those that keep lot sizes large, require large homes to be built, constrain development relative to market demands, and limit lower-cost housing. SDC policy in such communities merely reflects fundamental orientations of planning policy. Exclusionary housing policies, per se, are likely not constitutional. On the other hand, some communities are naturally more expensive to accommodate considering terrain, land costs, and highly insensitive demand. With or without SDCs, those communities will be exclusionary.

Excess Benefits to Existing Residents

Arguments that SDCs by themselves generate excess benefits to existing residents are not yet founded. To the extent that the overall fiscal base of communities is improved because of SDCs, existing residents may enjoy additional benefits; but those benefits would accrue no matter how growth was accommodated. Indeed, it is quite possible that without accommodating growth through SDCs, growth would be frustrated and existing residents may not enjoy additional benefits. Moreover, the assertion that existing house values may rise because SDCs raise the price of new housing is also misplaced. SDCs accommodate development by expanding supply of buildable land and may actually dampen price effects on both new and existing housing. Housing prices of existing homes may actually climb higher in the absence of SDCs.

Intergenerational Effects

Intergenerational effects (where one generation of residents make another generation better or worse off) only occur if SDCs are associated with increasing the price of new housing and the price of comparable housing also rises. This can only occur where markets are relatively insensitive to changes in house prices. In these situations, it is probable that SDCs by themselves would not raise the price of housing. Rather, housing prices would likely rise anyway in order to capture the greatest prices possible. Indeed, it is possible that with SDCs aiding in increasing the supply of buildable land, the price effects are dampened. If there are "windfall" profits, they are substantially taxed away through property, real estate transfer, and capital gains taxes. Again, however, the windfalls are likely attributable to a shift in community household composition that would occur anyway. Finally, consider that SDCs are relatively small in most communities. They rarely go above about 5% of total house price, or something less than real estate commissions. Unlike real estate commissions, however, SDCs go to pay for the very facilities that make facilities possible.

Political Representation

SDCs certainly affect new buyers of houses and new tenants of nonresidential developments, and it is possible that people who pay SDCs are not represented in the policy adopting system development charges. However, the people typically writing checks out for SDCs are developers and, in most communities, developers have greater than proportionate influence on local policy-making. Moreover, a large share of new housing is purchased not by new residents, but by existing residents who are moving up. Given those two considerations, it could be argued that representation of future fee payers during SDC policy-making is at least adequate if not weighted toward those residents.

Public Responsibility

Finally, SDC policy requires planning, facility programming, and a rigorous financial plan for buying new infrastructure. Responsible actions by public officials would do these things as a matter of course. However, SDCs are the glue that holds the entirety of development policy together. Moreover, when properly designed to account for substantially different costs of accommodating development in different parts of the community or with different facilities, SDC design comes close to achieving efficient and equitable outcomes. One could argue that by not using system development charges, communities may be shirking public responsibility.

SUMMARY

The proposition is advanced that system development charge policy, when considered in the context of all the planning, programming, and financial rigor preceding it, is a win-win policy in terms of public finance criteria, development efficiency, and equity and ethical implications.

6   Alternative Methods of Calculating System Development Charges

INTRODUCTION

This chapter presents several ways in which to calculate system development charges (SDCs) for water, wastewater, and stormwater facilities. It will also suggest which approaches are appropriate under certain conditions, and which are not. Indeed, the author has identified at least eight calculation approaches used around the U.S., all of which are reviewed in this chapter. They include:

- Market capacity method
- Prototypical system method
- Growth-related cost allocation method
- Recoupment value method, also known as the buy-in method
- Replacement cost method
- Marginal cost method
- Average cost method
- Systemwide and growth-related cost-attribution method

The first two methods are not recommended because of important shortcomings in meeting *rational nexus* criteria: the market capacity method and the prototypical system method. The market capacity method is principally based on "what the market is willing to bear". An example illustrates its limitations. Consider the case of a developer who received approval for a 70-unit residential subdivision in a West Coast city in 1979 where the water system connection fee at the time of approval was $210. Because of its isolation and small size, the City Council believed that the entire demand for new residential units for the next 5 to 10 years would be absorbed by this single subdivision. The day after the subdivision was approved, the city raised its water connection fee by $1000 to $1210. The developer's consultant calculated that the appropriate fee should be less than $300 and perhaps the original $210. The consultant recommended litigation, but the developer — who depended on good relations with the city as both a developer and resident, and who determined that legal fees would exceed the savings even if he won — decided to pay up. This approach, unfortunately used extensively across the U.S., fails to meet *rational nexus* criteria and is not recommended.

The prototypical system method involves calculating SDCs on the basis of a system fully built-out in a comparable community. While this may seem reasonable, there are two major shortcomings. First, no two communities are alike in their development patterns, timing of development, terrain, or policies with respect to distributing costs as they occur. Second, there is no assurance that the costs incurred by one community will equal costs of another. The problem is that by charging SDCs in one community based on a prototypical system, the SDCs may exceed the actual costs of the community in installing a system, which would be inconsistent with *rational nexus* criteria.

There are six methodologies that are appropriate to apply to most communities under particular conditions. These will now be reviewed.

Basic Assumptions

To best understand the differences among the SDC calculation methods, it is important to compare SDC outcomes with a common set of assumptions. For any given system, these categories of assumptions are known or can be readily determined. Suppose that for one community, these assumptions have characteristics as shown in Table 1 as they apply to a wastewater system. The meaning of each assumption is described below.

System Capacity, Gallons, 1995— This is the community's capacity to treat wastewater on an average daily or peak daily basis for the current year.

Capacity Before 1991 Expansions, Gallons — Suppose the community embarked on a program to expand the current system to accommodate new development in 1991. Prior to such expansion, the community had a level of capacity that served existing development although its preexpansion capacity may have included some excess

Table 1 Alternative Wastewater System Development Charge Methods — Basic Assumptions

Assumptions	Figure
System capacity, gallons, 1995	42,000,000
Capacity before 1991 expansions, gallons	40,000,000
Recent & planned expansions, 1991–1999, gallons	8,000,000
Total existing & planned capacity	48,000,000
Capacity used, 1991	24,700,000
Excess capacity, 1991	15,300,000
ERU factor, gallons	246
New development demand, gallons, 1995–1999	5,100,000
Total replacement cost	$138,754,774
Systemwide replacement cost	$126,536,568
Growth-related replacement cost, since 1991	$12,218,206
Total fixed asset value, 1995	$105,938,385
Systemwide fixed asset value	$93,966,655
Growth-related fixed asset value, since 1991	$11,971,730
CIP improvements, 1995 through 1999	$194,180,655
Systemwide CIP improvements, 1995–1999	$25,275,000
CIP Growth-related improvements, 1995–1999	$133,334,907

capacity able to accommodate new development. The year 1991 can be considered the base year for calculating SDCs under some methods described below.

Recent and Planned Expansions, 1991–1999, Gallons — Suppose the community has a long range plan to expand capacity to some level between 1991 and 1999, after which it has no plans for further expansion. This is a kind of "build out" scenario, although the network accessing this build-out capacity may take several years beyond 1999 to install.

Total Existing and Planned Capacity — This is the total capacity anticipated to be in place at the end of an expansion period, which is 1999 in this example.

Capacity Used, 1991 — This is the average or peak daily demand during the base year.

Excess Capacity, 1991 — This is the difference between base year capacity (before expansion) and average or peak daily demand during the base year. This is the amount of treatment capacity since system expansion to accommodate new development.

ERU Factor, Gallons — This is the equivalent residential unit factor that will be used to calculate SDCs by ERU for each method.

New Development Demand, Gallons, 1995–1999 — This is an estimate of the change in system demand attributed to new development during the period of the capital improvements program (CIP) during 1995 and 1999.

Total Replacement Cost — This is the amount of money required to replace the entire system that is in currently in place, irrespective of how the system was financed in the past. In this example, it is assumed that only local financing has been used to finance the system and its planned expansions.

Systemwide Replacement Cost — This is the cost of replacing elements of the system that serve all development, whether new or existing.

Growth-Related Replacement Cost — This is the cost of replacing elements of the system that substantially serve only new development.

Total Fixed Asset Value, 1995 — This is the replacement cost of the entire system, less accumulated depreciation. It is called "equity".

Systemwide Fixed Asset Value — This is the fixed asset value of that share of the system that serves all development, whether new or existing.

Growth-Related Fixed Asset Value, Since 1991 — This is the fixed asset value of that share of the system that substantially serves only new development constructed since 1991, the base year of analysis for some SDC calculation methods. This would include, for example, expansions to wastewater treatment installed between 1991 and the current year.

CIP Improvements, 1995–1999 — This is the total cost of installing improvements to accommodate existing and anticipated development during a typical 5-year capital improvement program (CIP), from 1995 to 1999.

Systemwide CIP Improvements, 1995–1999 — This is the share of CIP improvements that serves all development, whether new or existing. An example may be upgrades to treatment systems to comply with new environmental regulations or replacing a pump station installed 30 years ago.

CIP Growth-Related Improvements, 1995–1999 — This is the share of CIP improvements that substantially serves new development.

Table 2 Growth-Related Cost Allocation Method

Calculation considerations	Results
CIP improvements, 1995–1999	$194,180,655
New development demand, gallons, 1995–1999	5,100,000
Cost allocation cost per gallon	$38.07
ERU factor, gallons	246
Growth-related bases SDC per ERU	$9365.22

This would include, for example, planned expansions to wastewater treatment.

Growth-Related Cost Allocation Method

The growth-related cost allocation method, presented in Table 2, applies all costs reflected in the CIP to new growth expected to occur during the CIP period. The principal advantage of this method is its simplicity, but there are important pitfalls. If the CIP includes considerable expansion of capacity that will benefit users beyond 1999, the entire cost of such expansion is borne only by new connections occurring between 1995 and 1999. The SDCs would be too high and violate *rational nexus* principles because the capacity needed to accommodate new development is less than the capacity new development financed. Alternatively, if the CIP includes no major expansions, in part because prior CIPs financed such expansions, new development is paying less than its proportionate share of the system capacity it uses.

There are two circumstances under which the growth-related method is appropriate. One is when the system being installed is a complete package and the period of time over which costs are related to growth is the build-out period. the second is when, because of the bumpy nature of capacity expansion, the increment of capacity added exceeds the projected demand over a typical capital financing period.

Recoupment Value Method

The recoupment value method (see Table 3) essentially results in new development reimbursing existing development for new development's proportionate share of the cost of existing improvements based on total system treatment capacity available in the current year. Recoupment is

Table 3 Recoupment Value Method

Calculation consideration	Result
Total fixed asset value, 1995	$105,938,385
System capacity, gallons, 1995	42,000,000
Recoupment value per gallon	$2.52
ERU factor, gallons	246
Recoupment based SDC per ERU	$620.50

based only on the fixed asset value of the entire system. It does not distinguish between improvements made mostly for the benefit of new development, nor does it consider the cost of expanding system capacity to accommodate new development. As such, the recoupment approach will usually result in SDCc lower than the cost of accommodating new development. Moreover, this method does not consider the financing costs incurred by existing development to make excess capacity available to new development.

This method is popular among developers, because they pay considerably lower fees. Other methods are more accurate in reflecting the real costs of accommodating development. It may also be popular among elected officials who see the recoupment fee as a way to keep costs low on "economic" development. However, it will be the existing taxpayers and ratepayers who finance much, if not all, of the real costs of accommodating development. Moreover, this method does not consider the financing costs incurred by existing development to make excess capacity available to new development.

About the only circumstance under which this approach reasonably reflects the cost of accommodating new development is when the system has been completely built-out, possesses substantial excess capacity to accommodate new development on an in-fill basis, is in no need of major system upgrades, and existing taxpayers through their public officials are willing to forego reimbursement of the financing costs incurred in making excess capacity available to new development.

Replacement Cost Method

The replacement cost method (see Table 4) is conceptually similar to the recoupment value method, with the only difference being that it is based on the cost of replacing the entire system presently in place.

Table 4 Replacement Cost Method

Calculation consideration	Result
Total replacement cost	$138,754,774
System capacity, gallons, 1995	42,000,000
Replacement value per gallon, 1995	$3.30
ERU factor, gallons	246
Replacement cost based SDC per ERU	$812.71

The result is higher SDCs than calculated under the recoupment value method; otherwise, it suffers from the same limitations. This method is appropriate when the system has been completely built-out, or possesses substantial excess capacity to accommodate new development on an in-fill basis, is in no need of major system upgrades, and public officials wish to recoup financing costs previously incurred. Under this scenario, the SDC reasonably reflects the costs of providing service to new development on an incremental or marginal basis. If expansion costs are in fact reasonably close to replacement costs on a per-unit basis, this approach is quite reasonable.

Marginal Cost Method

Marginal cost is defined as the cost of providing the next unit of development. The replacement value method is a form of marginal cost pricing, but only when expansion costs to accommodate new development are reasonably close to replacement costs on a per-unit-of-demand basis.

Marginal cost is usually considered the cost per unit of expanding system capacity and the system network on a per-unit basis to accommodate new development. There are two practical limitations with this approach. First, while the cost to expand treatment capacity to accommodate new development may be fairly easily determined, what is less certain is the cost of providing the network by which new development accesses treatment. Second, the approach does not usually consider facilities installed in the past at lower real costs that may also be used by new development.

Marginal cost is defined here as composed of two parts resulting in *growth-related* marginal costs. The first part is the replacement cost of existing growth-related facilities installed since the base year,

Table 5 Marginal Cost Method

Calculation consideration	Amount
Growth-related replacement cost, since 1991	$12,218,206
CIP growth-related improvements, 1994–1999	$133,334,907
Total growth-related marginal cost	$145,553,113
Recent & planned expansions, 1991–1999, gallons	8,000,000
Growth-related marginal cost/gallon	$18.19
ERU factor, gallons	246
Marginal cost based SDC per ERU	$4474.74

1991. These are facilities that have been installed in the recent past to serve new development, but adjusted to reflect inflation. The second part is the cost of installing CIP growth-related facilities. These two figures are summed and then divided by total treatment capacity added to the system since the base year, 1991which is shown in Table 5.

Notice that replacement cost is used and not asset value, which includes depreciation. Marginal cost analysis is concerned with the cost of serving the *next* unit of demand. Well-designed capital improvement programs provide continuous replacement and upgrading of facilities to maintain their value to the system.

There are important shortcomings to strict application of marginal cost analysis. For example, it does not consider the role of existing excess capacity in serving new development; if the costs of providing excess capacity are substantially less than the costs of expanding capacity, it is possible that new development would be paying more than its proportionate share of the cost of service actually received. As such, as new development absorbs excess capacity that was purchased at lower cost than new capacity, total revenues generated will exceed total costs. This would be inconsistent with *rational nexus* criteria. In addition, this approach does not account for the possibility that other elements of the system not reflected in these particular marginal cost calculations may nonetheless accommodate new development.

The marginal cost method is appropriate to those situations where there is little or no excess capacity, new development is not likely to use existing facilities that serve existing development, and the CIP reflects the total costs of serving new development using

Table 6 Average Cost Method

Calculation consideration	Amount
Total replacement cost	$138,754,774
Total CIP expenditures	$194,180,655
Total costs	$332,935,429
Total existing and planned capacity, 1999, gallons	48,000,000
Average cost/gallon	$6.94
ERU factor, gallons	246
Average cost based SDC per ERU	$1706.29

expanded treatment capacity. These conditions are not likely to be found in many communities.

Average Cost Method

The average cost method (see Table 6) is like the marginal cost method except that instead of allocating costs to new development, the costs of replacing and expanding the entire system are considered in relation to the total capacity of the system to accommodate development, both existing and new. While it may appear to address some of the shortcomings of the marginal cost method reviewed above, it can have the effect of understating the costs of accommodating new development if new treatment capacity costs are considerably more than the cost of replacing existing capacity. On the other hand, if expansion costs are lower than replacing existing capacity, more revenue would be collected than needed to accommodate new development.

As in the case of marginal cost analysis, average cost analysis is based on replacement or expansion costs, not asset values that include depreciation.

The average cost approach is suitable only when the costs of expansion are similar to the costs of replacing the existing system on a per-unit basis. It is more often the case that average costs are less than the costs of accommodating new development; thus, revenues received would not equal revenues needed.

Total Cost Attribution Method

The total cost attribution method (see Table 7) considers both the contribution of systemwide facilities and growth-related facilities to

Table 7 Total Cost Attribution Method

Calculation consideration	Amount
Systemwide fixed asset value	$105,938,385
Systemwide CIP improvements, 1994–1999	$60,845,748
Total systemwide value	$166,784,133
Preexpansion capacity, gallons	48,000,000
Systemwide value and cost per gallon	$3.47
Growth-related asset value	$11,971,730
CIP growth-related improvements, 1994–1999	$133,334,907
Total growth-related asset value and CIP improvements	$145,306,637
Total capacity for new development since 1991, gallons	$23,300,000
CIP improvements per gallon based on excess capacity	$6.24
Asset value and CIP improvements per gallon	$9.71
ERU factor, gallons	246
Total cost attribution based SDC per ERU	$2388.91

Table 8 Summary of Methods

Method	SDC/ERU
Growth-related cost allocation method	$9365.22
Recoupment value method	$620.50
Replacement cost method	$812.71
Marginal cost method	$4474.74
Average cost method	$1706.29
Total cost attribution method	$2388.91

the accommodation of new development. This has the advantage of accounting for the impact of new development on systemwide facilities that serve both existing and new development. The method also allocates growth-related asset value and CIP improvements to capacity expansion as well as excess capacity. The method recognizes that delivery of excess capacity often requires extension of lines, installation of pumps, and construction of related facilities to accommodate new development. Asset value is used because CIP improvements (which include replacement and rehabilitation of existing facilities) will, over time, offset depreciation of those assets.

The total cost allocation method is perhaps the most suitable approach to calculating SDCs. It combines all the elements of the *rational nexus* criteria, including such *Banberry* factors as how existing facilities were financed, the costs of accommodating new development, and the time-value of money. However, modifications of the method may be necessary, such as substituting systemwide asset value and growth-related asset value with systemwide replacement cost and growth-related replacement cost.

SUMMARY

Table 8 summarizes these six SDC calculation methods. Each may be appropriate under particular circumstances. However, the total cost allocation method is probably the most suitable method to most systems in use today.

7   Calculating System Development Charges for Water and Wastewater Facilities Based on Total Cost Attribution

INTRODUCTION

This chapter applies the principles of Chapters 2 and 3 to the calculation of system development charges (SDCs) for water and wastewater systems for a hypothetical city — the City of Columbia, (data for which are a composite of separate studies). The SDC methodology chosen is the "total cost attribution" approach described in Chapter 6 because it is the method having perhaps the largest number of practical applications.

Consistent with the general outline for calculating SDCs reviewed in Chapter 3, this example will consider:

- Service area
- Level of service and projections of demand
- Inventory
- Assessment
- Facility expansion costs
- Gross system development charge
- Debt service costs and credits
- Net system development charge
- Alternative SDC schedules

SERVICE AREA

In this example, the Columbia Water Works is a water utility originally created by an act of the State General Assembly in 1902 and extended under the Charter of Columbia in 1921. The Water Works assumed responsibility for the sanitary sewer system in 1956. The operation and management of Columbia Water Works is under the jurisdiction of the Board of Water Commissioners, consisting of the Mayor of Columbia and four members appointed to staggered 4-year terms by the Council of Columbia.

Columbia Water Works supplies water to approximately 62,000 metered water customers and sewage collection and treatment facilities for approximately 95% of the population of Columbia. To provide those services, it operates a water production and treatment plant and a wastewater treatment plant, installs water storage, transmission and distribution facilities, wastewater collection lines, and pumping facilities, and operates a service center that supports the maintenance of the system.

Columbia Water Works serves all of the city and the county within which it is situated. It has two water meters located on the county line through which it provides wholesale water service to the adjacent county, but these sales (approximately 52,500 gpd in 1992) account for less than 0.2 of 1% of its average daily water sales.

This example treats the entire jurisdiction of Columbia Water Works as a single systemwide service area for both water and wastewater service. The use of a systemwide service area for water and wastewater capital expansion charges is justified in several respects. First, from the perspective of the individual customer, the location of treatment plant, size and placement of lines, method of wastewater disposal, etc. are discretionary decisions made by the utility. Equal quality service is provided to all customers. Second, the water and wastewater systems are integrated in such a manner, such as through redundant water transmission facilities and inter-basin sewage pumpovers, that it is difficult to allocate facilities to specific subareas. Third, these systems contain central facilities, such as Columbia's water and wastewater treatment plants, that serve the entire jurisdiction and cannot be allocated to a specific subarea. Finally, all ratepayers are bound by bond and loan covenants to retire debt service incurred on behalf of existing and future ratepayers regardless of their location.

LEVEL OF SERVICE AND PROJECTIONS OF DEMAND

For water and wastewater facilities, the most important level of service component is the central treatment plants. Treatment plants are designed to accommodate peak daily demand, but their permitted capacity is based on the average daily demand that they can accommodate. The capacity of pumps, lines, storage tanks, and facilities other than treatment plants is only of secondary importance. The Columbia water treatment plant, for example, can treat up to 67.5 million gallons per day (MGD), but its permitted capacity is rated at an average daily demand of 55 MGD, calculated on a monthly basis.

Columbia has established levels of service in terms of *equivalent residential units* (ERU) based on a typical $^5/_8$-in. residential water meter connection. For water consumption, the city's level of service standard is based an average annual flow of 313 gallons per day (gpd). This figure is based on an average daily flow per $^5/_8$-in. meter in 1990. It is possible that this figure may be reduced because of conservation and improvements in the system; but until this separate determination is made, the level of service standard will be 313 gpd/ERU. The city has established the level of service for wastewater at 80% of the water-based ERU level, or 250 gpd for detached residential and 955 return, or 297 gpd, for nonresidential and attached residential.

Using these levels of service as a measure of demand, it appears that the wastewater and water treatment plants have sufficient capacity to meeting existing demand, as well as projected demand for about 15 years (see Table 1). In 1992, the water treatment plant had an average daily demand of 32.7 MGD, which was less that two thirds of its permitted capacity of 55 MGD. The wastewater treatment plant had an average daily flow of about 27.2 MGD, which was about two thirds of its permitted capacity of 42 MGD. Historical demand and supply figures for 1990 to 1992, and projected supply and demand for the next 2 decades are shown in Table 1. These data will be referred to throughout this example.

INVENTORY OF EXISTING FACILITIES

The inventory of existing facilities is necessary to determine the extent to which new development is benefiting from investments

Table 1 Wastewater/Water System Demand and Capacity

Year	Wastewater facilities		Water facilities	
	Demand (MGD)	Capacity (MGD)	Demand (MGD)	Capacity (MGD)
1985	24.3	40.0	25.8	55.0
1986	25.5	40.0	26.4	55.0
1987	26.6	40.0	27.8	55.0
1988	27.1	40.0	28.6	55.0
1989	28.2	40.0	29.7	55.0
1990	29.0	40.0	30.5	55.0
1991	24.7	42.0	29.9	55.0
1992	27.2	42.0	32.7	55.0
1993	28.2	42.0	35.3	55.0
1994	29.0	42.0	36.3	55.0
1995	29.8	42.0	37.3	55.0
1996	30.6	42.0	38.3	55.0
1997	31.5	42.0	39.4	55.0
1998	32.4	42.0	40.5	55.0
1999	33.3	48.0	41.6	55.0
2000	34.3	48.0	42.9	55.0
2001	35.3	48.0	44.1	55.0
2002	36.3	48.0	45.4	55.0
2003	37.3	48.0	46.6	55.0
2004	38.4	48.0	48.0	55.0
2005	39.5	48.0	49.4	55.0
2006	40.6	48.0	50.8	55.0
2007	41.8	48.0	52.3	55.0
2008	43.0	48.0	53.8	55.0
2009	44.2	48.0	55.3	55.0
2010	45.5	48.0	56.9	55.0
2011	46.8	48.0	58.5	55.0
2012	48.0	48.0	60.0	55.0
2013	48.9	NA	61.1	55.0
2014	50.0	NA	62.5	55.0
2015	51.1	NA	63.9	55.0
2016	52.2	NA	65.3	55.0
2017	53.3	NA	66.6	55.0
2018	54.4	NA	68.0	55.0
2019	55.5	NA	69.4	55.0

Table 1 Wastewater/Water System Demand and Capacity
(Continued)

Year	Wastewater facilities		Water facilities	
	Demand (MGD)	Capacity (MGD)	Demand (MGD)	Capacity (MGD)
2020	56.6	NA	70.8	55.0
2021	57.7	NA	72.1	55.0
2022	58.8	NA	73.5	55.0

made by current users and, further, the extent to which such benefits may be recouped in part through SDCs. There are two general ways in which an inventory may be assembled. The first is presented here. It involves identifying only those facilities for which local resources were used to acquire, replace, or upgrade system components. The second, which is not presented here, would inventory *all* facilities, including those provided by developers. Since developer contributions were not financed by local rate payers, such facilities are excluded from this step in the analysis.

The essential elements of the inventory are shown in Table 2A for wastewater facilities and in Table 2B for water facilities. To the analyst, these tables present three very important pieces of information relevant to the calculation of *rational nexus* SDCs. They are current asset value of locally financed facilities, current asset value of systemwide improvements locally financed, and current asset value of growth-related improvements locally financed. These distinctions are important for different methods of SDC calculation, which were reviewed in Chapter 6. The meaning of each column is described here.

Facility Name/Location

This is the list of components of the system. Only those facilities that have involved at least some local funds are inventoried. (Actually, it is recommended that a complete inventory be developed that includes developer-installed or -financed facilities, as well as facilities financed totally by other governmental agencies; but for SDC calculation purposes, only those facilities that have some local contribution would be listed in this inventory.)

Table 2A Inventory and Current Value of Wastewater Facilities

No.	Facility Name/location	Year built	Useful life (yrs)	Total amount paid ($)	Amount locally paid ($)	ENR factor 1992	Local replacement value 1992 ($)	Deprec percent 1992 (%)	Current local asset value 1992 ($)	Percent system improve-ment (%)	Percent growth related (%)	Current value local systemwide improvements ($)	Current value local growth related improvements ($)
1	System network expansion	1960	50	1,237,658	1,237,658	6.0878	7,534,612	64.00	2,712,460	100	0	2,712,460	0
2	System network expansion	1964	50	53,193	53,193	5.3365	283,864	56.00	124,900	100	0	124,900	0
3	System network expansion	1965	50	3,270,356	3,270,356	5.1204	16,745,531	54.00	7,702,944	100	0	7,702,944	0
4	Sewer plant expansion	1965	50	2,268,728	567,182	5.1204	2,904,199	54.00	1,335,932	100	0	1,335,932	0
5	System network expansion	1966	50	3,600	3,600	4.8926	17,613	52.00	8,454	100	0	8,454	0
6	System network expansion	1967	50	2,882,865	720,716	4.6075	3,320,700	50.00	1,660,350	100	0	1,660,350	0
7	Sewer plant expansion	1967	50	503,521	125,880	4.6075	579,993	50.00	289,997	100	0	289,997	0
8	System network expansion	1968	50	4,600	4,600	4.2123	19,377	48.00	10,076	100	0	10,076	0
9	System network expansion	1969	50	945,888	236,472	3.8766	916,707	46.00	495,022	100	0	495,022	0
10	Sewer plant expansion	1969	50	232,318	58,080	3.8766	225,151	46.00	121,582	100	0	121,582	0
11	System network expansion	1970	50	1,525,920	1,525,920	3.5010	5,342,246	44.00	2,991,658	100	0	2,991,658	0
12	Sewer plant expansion	1970	50	137,603	137,603	3.5010	481,748	44.00	269,779	100	0	269,779	0
13	System network expansion	1974	50	482,806	482,806	2.4079	1,162,549	36.00	744,031	100	0	744,031	0
14	Contr–Osbrn–Wlmsn	1974	50	2,645,451	2,645,451	2.4079	6,369,981	36.00	4,076,788	100	0	4,076,788	0
15	Line ext & renews	1974	50	106,056	106,056	2.4079	255,372	36.00	163,438	100	0	163,438	0
16	N Cols sewers cont	1974	50	29,593	29,593	2.4079	71,257	36.00	45,604	100	0	45,604	0
17	Metro contract	1974	50	27,553	27,553	2.4079	66,345	36.00	42,461	100	0	42,461	0
18	2 Submersbl pumps	1974	35	3,446	3,446	2.4079	8,298	51.43	4,030	100	0	4,030	0
19	Wiring/lift sta	1974	35	259	259	2.4079	624	51.43	303	100	0	303	0
20	6 x 4 w/20 hp motor	1974	35	2,304	2,304	2.4079	5,548	51.43	2,695	100	0	2,695	0
21	Willowbend sewers	1975	50	27,037	27,037	2.2024	59,546	34.00	39,300	100	0	39,300	0
22	Line ext & renewal	1975	50	86,341	86,341	2.2024	190,157	34.00	125,504	100	0	125,504	0
23	Secondary treatment	1975	50	6,754,671	1,688,668	2.2024	3,719,122	34.00	2,454,621	100	0	2,454,621	0
24	Lindsey Ck Cont B	1976	50	890,199	0	2.0317	0	32.00	0	50	0	0	0
25	Contract 1 sew sys	1976	50	328,452	328,452	2.0317	667,316	32.00	453,775	100	0	453,775	0
26	Contract 2 sew sys	1976	50	403,295	403,295	2.0317	819,374	32.00	557,174	100	0	557,174	0
27	Contract 3 sew sys	1976	50	313,269	313,269	2.0317	636,469	32.00	432,799	100	0	432,799	0
28	201 Planning study	1976	50	16,742	16,742	2.0317	34,015	32.00	23,130	100	0	23,130	0
29	Col Mus facity 100	1976	50	123,258	123,258	2.0317	250,423	32.00	170,288	100	0	170,288	0
30	Sewer line ext & ren	1976	50	63,145	63,145	2.0317	128,292	32.00	87,239	100	0	87,239	0
31	Sewer plant replacements	1976	50	8,110	8,110	2.0317	16,477	32.00	11,204	100	0	11,204	0

ID	Description	Year											
32	Air cond	1976	35	2,187	2,187	2.0317	4,443	45.71	2,412	100	0	2,412	0
33	Routine line extensions	1977	50	75,113	75,113	1.9019	142,857	30.00	100,000	100	0	100,000	0
34	Fiygt sub pump	1977	35	6,452	6,452	1.9019	12,271	42.86	7,012	100	0	7,012	0
35	Cap Cont Div I	1978	50	1,747,274	1,747,274	1.7633	3,080,968	28.00	2,218,297	100	0	2,218,297	0
36	Cap Cont A Div II	1978	50	1,344,494	1,344,494	1.7633	2,370,746	28.00	1,706,937	100	0	1,706,937	0
37	Cap Cooper Ck Intr	1978	50	841,616	841,616	1.7633	1,484,021	28.00	1,068,495	100	0	1,068,495	0
38	Routine line ext	1978	50	25,631	25,631	1.7633	45,195	28.00	32,540	100	0	32,540	0
39	Cap Cp Con 5-7719	1978	50	435,333	435,333	1.7633	767,623	28.00	552,689	100	0	552,689	0
40	Swr improv cont #4	1978	50	337,202	337,202	1.7633	594,588	28.00	428,103	100	0	428,103	0
41	CSO #7811	1979	50	130,839	130,839	1.6111	210,795	26.00	155,988	100	0	155,988	0
42	Routine line ext	1979	50	24,860	24,860	1.6111	40,052	26.00	29,638	100	0	29,638	0
43	Completed construct	1979	50	178,781	178,781	1.6111	288,034	26.00	213,145	100	0	213,145	0
44	8 in. Flanged valve	1980	35	10,725	10,725	1.4985	16,071	34.29	10,560	100	0	10,560	0
45	Renewal ext aur\ent	1981	50	57,575	57,575	1.3691	78,826	22.00	61,484	100	0	61,484	0
46	Fence N sew pump sta	1981	35	3,650	3,650	1.3691	4,997	31.43	3,426	100	0	3,426	0
47	Repair pump sta	1981	35	24,516	24,516	1.3691	33,565	31.43	23,016	0	0	0	0
48	Repair motor	1981	35	18,889	18,889	1.3691	25,861	31.43	17,733	0	0	0	0
49	42 IN intercept 4St	1982	50	40,000	40,000	1.2808	51,232	20.00	40,986	100	0	40,986	0
50	Oliver pump sta	1982	35	400,533	400,533	1.2808	513,003	28.57	366,438	100	0	366,438	0
51	Captiz routin ln ext	1982	50	57,714	57,714	1.2808	73,920	20.00	59,136	100	0	59,136	0
52	Repair pump & valves	1982	35	13,480	13,480	1.2808	17,265	28.57	12,332	0	0	0	0
53	Reclamation project	1982	50	1,809,193	1,809,193	1.2808	2,317,214	20.00	1,853,771	100	0	1,853,771	0
54	Heif Ck plant	1982	50	386,349	386,349	1.2808	494,836	20.00	395,869	100	0	395,869	0
55	Meadowrun lift equip	1982	35	9,338	9,338	1.2808	11,960	28.57	8,543	100	0	8,543	0
56	Bull Ck sewer ext	1982	50	1,022,615	1,022,615	1.2808	1,309,765	20.00	1,047,812	100	0	1,047,812	0
57	Capitalize eng JJG	1983	50	80,753	80,753	1.2309	99,399	18.00	81,507	100	0	81,507	0
58	Sew sys-autid 5 83	1983	50	115,373	115,373	1.2309	142,013	18.00	116,451	100	0	116,451	0
59	Col pipe-I185 sewer	1983	50	220,257	220,257	1.2309	271,114	18.00	222,313	100	0	222,313	0
60	Lynch Rd outfal sr	1983	50	1,309,878	1,309,878	1.2309	1,612,329	18.00	1,322,110	100	0	1,322,110	0
61	Audit cap sew ext	1983	50	129,130	129,130	1.2309	158,946	18.00	130,336	100	0	130,336	0
62	Capitalize eng JJG	1983	50	28,101	28,101	1.2309	34,590	18.00	28,364	100	0	28,364	0
63	ABS submersible	1984	35	12,340	12,340	1.2208	15,065	22.86	11,621	100	0	11,621	0
64	Angus double hose	1984	35	6,531	6,531	1.2208	7,973	22.86	6,150	100	0	6,150	0
65	2N Ave sta frce mn	1984	50	2,005,778	2,005,778	1.2208	2,448,654	16.00	2,056,869	100	0	2,056,869	0
66	Impella shafts	1984	35	3,764	3,764	1.2208	4,595	22.86	3,545	100	0	3,545	0
67	ABS pump sta	1984	35	10,906	10,906	1.2208	13,314	22.86	10,270	100	0	10,270	0
68	ABS pumping stn	1985	35	24,514	24,514	1.1965	29,331	20.00	23,465	100	0	23,465	0

Table 2A Inventory and Current Value of Wastewater Facilities (Continued)

No.	Facility Name/location	Year built	Useful life (yrs)	Total amount paid ($)	Amount locally paid ($)	ENR factor 1992	Local replacement value 1992 ($)	Deprec percent 1992 (%)	Current local asset value 1992 ($)	Percent system improvement (%)	Percent growth related (%)	Current value local systemwide improvements ($)	Current value local growth related improvements ($)
69	3 ABS sub pump AFP	1985	35	39,641	39,641	1.1965	47,430	20.00	37,944	100	0	37,944	0
70	Sew sys audit 21 85	1985	50	248,528	248,528	1.1965	297,364	14.00	255,733	100	0	255,733	0
71	1 Teel pump	1986	35	1,450	1,450	1.1627	1,686	17.14	1,397	100	0	1,397	0
72	Grls marina pump	1986	35	27,959	27,959	1.1627	32,508	17.14	26,936	100	0	26,936	0
73	Am Legion temp sta	1986	50	69,976	69,976	1.1627	81,361	12.00	71,598	100	0	71,598	0
74	Am Legion 54 sew	1986	50	9,288	9,288	1.1627	10,799	12.00	9,503	100	0	9,503	0
75	4 × 4 × 7 Pump w/40 hp	1986	35	3,775	3,775	1.1627	4,389	17.14	3,637	100	0	3,637	0
76	Audit 23–32–40	1986	50	200,428	200,428	1.1627	233,038	12.00	205,073	100	0	205,073	0
77	4 20 hp pumps	1986	35	3,296	3,296	1.1627	3,832	17.14	3,175	100	0	3,175	0
78	4 20 hp pumps	1986	35	3,296	3,296	1.1627	3,832	17.14	3,175	100	0	3,175	0
79	Baker lift sta	1987	35	40,781	40,781	1.1297	46,070	14.29	39,487	100	0	39,487	0
80	Miller Rd lift sta	1987	35	73,917	73,917	1.1297	83,504	14.29	71,571	100	0	71,571	0
81	Electrogage contrl	1987	35	5,536	5,536	1.1297	6,254	14.29	5,360	100	0	5,360	0
82	Lift stations	1987	35	6,688	6,688	1.1297	7,555	14.29	6,475	100	0	6,475	0
83	Electrogage contrl	1987	35	6,089	6,089	1.1297	6,879	14.29	5,896	100	0	5,896	0
84	Shaw N bypass reloc	1987	50	16,377	16,377	1.1297	18,501	10.00	16,651	100	0	16,651	0
85	Flang plug valve	1987	35	9,548	9,548	1.1297	10,786	14.29	9,245	100	0	9,245	0
86	4 Duplex pump ctrl	1987	35	32,863	32,863	1.1297	37,125	14.29	31,820	100	0	31,820	0
87	Hardaway pump sta	1987	35	28,788	28,788	1.1297	32,522	14.29	27,875	100	0	27,875	0
88	Pipe liner Macon Rd	1987	50	122,714	122,714	1.1297	138,630	10.00	124,767	100	0	124,767	0
89	Duplex pump ctrl	1987	35	9,026	9,026	1.1297	10,197	14.29	8,740	100	0	8,740	0
90	Hardaway lift sta.	1987	35	6,647	6,647	1.1297	7,509	14.29	6,436	100	0	6,436	0
91	Green Island marina	1987	50	1,153	1,153	1.1297	1,303	10.00	1,173	100	0	1,173	0
92	Duplex AF30–4–4	1987	35	29,081	29,081	1.1297	32,853	14.29	28,158	100	0	28,158	0
93	Quail Ck lift stn	1987	35	12,066	12,066	1.1297	13,631	14.29	11,683	100	0	11,683	0
94	ABS pump accessors	1987	35	5,519	5,519	1.1297	6,235	14.29	5,344	100	0	5,344	0
95	Pump ctrls & valves	1987	35	22,246	22,246	1.1297	25,131	14.29	21,540	100	0	21,540	0
96	Audit entry 22B	1987	50	63,995	63,995	1.1297	72,295	10.00	65,066	100	0	65,066	0
97	Audit entry 21B	1987	50	239,337	239,337	1.1297	270,379	10.00	243,341	100	0	243,341	0
98	Standing Boy Creek	1987	50	684,954	684,954	1.1297	773,793	10.00	696,414	100	0	696,414	0
99	Gr Is Marina St	1987	50	498,294	498,294	1.1297	562,923	10.00	506,631	100	0	506,631	0

100	River interceptor	1987	50	1,074,590	1,074,590	1.1297	1,213,964	10.00	1,092,568	100	0	1,092,568	0
101	Valves parts	1987	35	2,872	2,872	1.1297	3,244	14.29	2,780	100	0	2,780	0
102	Plug valves	1987	35	8,176	8,176	1.1297	9,236	14.29	7,916	100	0	7,916	0
103	Intake drive chain	1987	35	6,137	6,137	1.1297	6,933	14.29	5,942	100	0	5,942	0
104	Intake drive chain	1987	35	6,137	6,137	1.1297	6,933	14.29	5,942	100	0	5,942	0
105	Sewer plant intake chain	1987	35	6,137	6,137	1.1297	6,933	14.29	5,942	100	0	5,942	0
106	Heat exchange	1987	35	9,737	9,737	1.1297	11,000	14.29	9,428	100	0	9,428	0
107	Op 10 software	1987	10	8,984	8,984	1.1297	10,149	50.00	5,075	100	0	5,075	0
108	Sluice gate parts	1987	50	4,905	4,905	1.1297	5,541	10.00	4,987	100	0	4,987	0
109	Sewer col. WPCP headwork	1987	50	682,223	682,223	1.1297	770,707	10.00	693,636	100	0	693,636	0
110	Auditn entry 18	1988	50	204,341	204,341	1.1034	225,470	8.00	207,432	100	0	207,432	0
111	JJG engineer fee	1989	50	7,306	7,306	1.0798	7,889	6.00	7,416	100	0	7,416	0
112	AJE#9 misc proj	1989	50	1,063,760	1,063,760	1.0798	1,148,648	6.00	1,079,729	100	0	1,079,729	0
113	River inter phase I	1989	50	2,942,709	2,942,709	1.0798	3,177,537	6.00	2,986,885	100	0	2,986,885	0
114	River inter phase I	1989	50	2,884,832	2,884,832	1.0798	3,115,042	6.00	2,928,139	100	0	2,928,139	0
115	River inter phase I	1989	50	407,927	407,927	1.0798	440,480	6.00	414,051	100	0	414,051	0
116	Polymate roof sys	1989	35	3,217	3,217	1.0798	3,474	8.57	3,176	100	0	3,176	0
117	JJG engineer fee	1989	50	1,500	1,500	1.0798	1,620	6.00	1,523	100	0	1,523	0
118	Phase I WPC plant	1989	50	12,588,574	12,588,574	1.0798	13,593,142	6.00	12,777,553	100	0	12,777,553	0
119	Phase I WPC eng fee	1989	50	889,332	889,332	1.0798	960,301	6.00	902,683	100	0	902,683	0
120	Cap routine line ext	1990	50	106,903	106,903	1.0590	113,210	4.00	108,682	100	0	108,682	0
121	AJE#8 routine line	1990	50	76,514	76,514	1.0590	81,028	4.00	77,787	100	0	77,787	0
122	Sewer col. feeder main	1990	50	2,664,797	2,664,797	1.0590	2,822,020	4.00	2,709,139	100	0	2,709,139	0
123	5 Wack hammers	1990	35	8,977	8,977	1.0590	9,507	5.71	8,964	100	0	8,964	0
124	Hotsey	1990	50	2,104	2,104	1.0590	2,228	4.00	2,139	100	0	2,139	0
125	So Col WWTF expans 42 MGD	1991	50	8,100,000	8,100,000	1.0348	8,381,880	2.00	8,214,242	100	100	0	8,214,242
126	Btl Frst WWTF conversion	1991	50	830,000	830,000	1.0348	858,884	2.00	841,706	100	100	0	841,706
127	Flat Rock Crk swr ext	1991	50	1,200,000	1,200,000	1.0348	1,241,760	2.00	1,216,925	100	100	0	1,216,925
128	Ind Prk sewer lines	1991	50	500,000	250,000	1.0348	258,700	2.00	253,526	100	100	0	253,526
129	1991 Bond issue costs	1991	20	68,000	68,000	1.0348	70,366	5.00	66,848	100	100	0	66,848
130	Heiferhorn Cr expansion	1991	50	1,421,192	1,421,192	1.0348	1,470,649	2.00	1,441,236	100	50	720,618	720,618
131	Exeter Court	1991	50	736,888	736,888	1.0348	762,532	2.00	747,281	100	0	747,281	0
132	Lake Oliver SS imp	1991	50	1,297,430	1,297,430	1.0348	1,342,581	2.00	1,315,729	50	100	0	657,865
133	Exeter Ct-repairs	1991	35	6,299	0	1.0348	0	2.86	0	0	0	0	0
134	4 8 87c pvmt brktrs	1991	35	4,192	4,192	1.0348	4,338	2.86	4,214	100	0	4,214	0
135	Welder power	1991	35	2,513	2,513	1.0348	2,600	2.86	2,526	100	0	2,526	0

Table 2A Inventory and Current Value of Wastewater Facilities (Continued)

No.	Facility Name/location	Year built	Useful life (yrs)	Total amount paid ($)	Amount locally paid ($)	ENR factor 1992	Local replacement value 1992 ($)	Deprec percent 1992 (%)	Current local asset value 1992 ($)	Percent system improvement (%)	Percent growth related (%)	Current value local systemwide improvements ($)	Current value local growth related improvements ($)
136	Welder power	1991	35	2,513	2,513	1.0348	2,600	2.86	2,526	100	0	2,526	0
137	Centrifuge	1991	35	1,882	1,882	1.0348	1,947	2.86	1,891	100	0	1,891	0
138	3 Vac press relief	1991	35	9,600	9,600	1.0348	9,934	2.86	9,650	100	0	9,650	0
139	33 Blower wheel	1991	35	3,965	3,965	1.0348	4,103	2.86	3,986	100	0	3,986	0
140	Vac-matic	1991	35	1,949	1,949	1.0348	2,017	2.86	1,959	100	0	1,959	0
141	GE magnebalaster	1991	35	19,992	19,992	1.0348	20,688	2.86	20,096	100	0	20,096	0
142	32 Pump silencers	1991	35	8,134	8,134	1.0348	8,417	2.86	8,176	100	0	8,176	0
143	24 Pump silencers	1991	35	5,224	5,224	1.0348	5,406	2.86	5,251	100	0	5,251	0
144	CSO project development	1991	50	4,000,000	4,000,000	1.0348	4,139,200	2.00	4,056,416	100	0	4,056,416	0
145	Wercoba Crk sewer-CSO	1991	50	3,000,000	3,000,000	1.0348	3,104,400	2.00	3,042,312	100	0	3,042,312	0
146	River interceptor/CSO	1991	50	9,500,000	9,500,000	1.0348	9,830,600	2.00	9,633,988	100	0	9,633,988	0
147	Bull Crk Crossing CSO	1991	50	500,000	500,000	1.0348	517,400	2.00	507,052	100	0	507,052	0
148	Bull Crk Basin I/I Study	1991	50	1,000,000	1,000,000	1.0348	1,034,800	2.00	1,014,104	100	0	1,014,104	0
149	SCADA projects	1991	50	750,000	375,000	1.0348	388,050	2.00	380,289	100	0	380,289	0
150	R Crk & Rsvlt Hts PS upgr	1991	35	135,000	135,000	1.0348	139,698	2.86	135,703	100	0	135,703	0
151	Steel muni boxes	1992	35	18,035	18,035	1.0000	18,035	0.00	18,035	100	0	18,035	0
152	2 Walker rammers	1992	35	4,190	4,190	1.0000	4,190	0.00	4,190	100	0	4,190	0
153	1976 Bond issue costs	1976	20	28,000	28,000	2.0317	56,888	80.00	11,378	100	0	11,378	0
154	1980 Bond issue costs	1980	25	33,000	33,000	1.4985	49,451	48.00	25,715	100	0	25,715	0
155	1985 Bond issue costs	1985	20	23,000	23,000	1.1965	27,519	35.00	17,887	100	0	17,887	0
156	Ins. value admin. prop	1992	NA	3,456,234	3,456,234	1.0000	3,456,234		3,456,234	100	0	3,456,234	0
	Total system values			103,183,100	91,470,608		139,502,754		106,649,330			93,966,655	11,971,730

Table 2B Inventory and Current Value of Water Facilities

No.	Facility Name/location	Year built	Useful life (yrs)	Total amount paid ($)	Amount locally paid ($)	ENR factor 1992	Local replacement value 1992 ($)	Deprec percent 1992 (%)	Current local asset value 1992 ($)	Percent system improvement (%)	Percent growth related (%)	Current value local systemwide improvements ($)	Current value local growth related improvements ($)
1	System expansion	1942	50	121,869	121,869	17.8134	2,170,901	100.00	0	100	0	0	0
2	System expansion	1943	50	118,162	118,162	17.1492	2,026,384	98.00	40,528	100	0	40,528	0
3	System expansion	1944	50	25,852	25,852	16.7517	433,065	96.00	17,323	100	0	17,323	0
4	System expansion	1945	50	166,058	166,058	16.1629	2,683,979	94.00	161,039	100	0	161,039	0
5	System expansion	1946	50	194,845	194,845	13.7477	2,678,671	92.00	214,294	100	0	214,294	0
6	System expansion	1947	50	264,421	264,421	11.7106	3,096,529	90.00	309,653	100	0	309,653	0
7	System expansion	1948	50	680,272	680,272	10.5837	7,199,795	88.00	863,975	100	0	863,975	0
8	System expansion	1949	50	735,921	735,921	10.5396	7,756,313	86.00	1,085,884	100	0	1,085,884	0
9	System expansion	1950	50	172,065	172,065	9.5273	1,639,315	84.00	262,290	100	0	262,290	0
10	System expansion	1951	50	151,269	151,269	9.2318	1,396,485	82.00	251,367	100	0	251,367	0
11	System expansion	1952	50	38,167	38,167	8.6037	328,377	80.00	65,675	100	0	65,675	0
12	System expansion	1953	50	78,342	78,342	8.2799	648,664	78.00	142,706	100	0	142,706	0
13	System expansion	1954	50	1,714,272	1,714,272	7.8678	13,487,549	76.00	3,237,012	100	0	3,237,012	0
14	System expansion	1955	50	64,835	64,835	7.5171	487,371	74.00	126,716	100	0	126,716	0
15	System expansion	1956	50	883,237	883,237	7.1861	6,347,029	72.00	1,777,168	100	0	1,777,168	0
16	System 1957	1957	50	100,000	100,000	6.8550	685,500	70.00	205,650	100	0	205,650	0
17	Filter plant	1958	50	731,019	731,019	6.5362	4,778,086	68.00	1,528,988	100	0	1,528,988	0
18	Tanks	1958	50	226,168	226,168	6.5362	1,478,279	68.00	473,049	100	0	473,049	0
19	System 1958	1958	50	1,269,261	1,269,261	6.5362	8,296,144	68.00	2,654,766	100	0	2,654,766	0
20	System 1959	1959	50	119,420	119,420	6.2534	746,781	66.00	253,906	100	0	253,906	0
21	Tanks	1960	50	1,548	1,548	6.0878	9,424	64.00	3,393	100	0	3,393	0
22	Filter plant	1960	50	650,312	650,312	6.0878	3,958,969	64.00	1,425,229	100	0	1,425,229	0
23	Filter plant	1961	50	76,390	76,390	5.9170	452,000	62.00	171,760	100	0	171,760	0
24	System 1962	1962	50	55,078	55,078	5.7423	316,274	60.00	126,510	100	0	126,510	0
25	Filter plant	1964	50	45,894	45,894	5.3365	244,913	56.00	107,762	100	0	107,762	0
26	System 1964	1964	50	37,373	37,373	5.3365	199,441	56.00	87,754	100	0	87,754	0
27	Filter plant	1967	50	1,495,959	1,495,959	4.6075	6,892,631	50.00	3,446,316	100	0	3,446,316	0
28	System 1967	1967	50	794,610	794,610	4.6075	3,661,166	50.00	1,830,583	100	0	1,830,583	0
29	Tanks	1969	50	2,081	2,081	3.8766	8,067	46.00	4,356	100	0	4,356	0
30	Filter plant	1969	50	88,881	88,881	3.8766	344,556	46.00	186,060	100	0	186,060	0
31	Tanks	1973	50	155	155	2.6091	404	38.00	250	100	0	250	0

Table 2B Inventory and Current Value of Water Facilities (Continued)

No.	Facility Name/location	Year built	Useful life (yrs)	Total amount paid ($)	Amount locally paid ($)	ENR factor 1992	Local replacement value 1992 ($)	Deprec percent 1992 (%)	Current local asset value 1992 ($)	Percent system improvement (%)	Percent growth related (%)	Current value local systemwide improvements ($)	Current value local growth related improvements ($)
32	System 1973	1973	50	362,822	362,822	2.6091	946,639	38.00	586,916	100	0	586,916	0
33	2 30 In valves	1974	35	19,860	19,860	2.4079	47,821	51.43	23,227	100	0	23,227	0
34	Line ext & renew	1974	50	129,591	129,591	2.4079	312,042	36.00	199,707	100	0	199,707	0
35	Tanks	1974	50	295,773	295,773	2.4079	712,192	36.00	455,803	100	0	455,803	0
36	System 1974	1974	50	996,421	996,421	2.4079	2,399,282	36.00	1,535,540	100	0	1,535,540	0
37	Alcon Assoc Contr	1974	50	2,173,227	2,173,227	2.4079	5,232,913	36.00	3,349,064	100	0	3,349,064	0
38	Legal Kelly-Champ	1974	50	1,310	1,310	2.4079	3,154	36.00	2,019	100	0	2,019	0
39	Addition line ext	1975	50	4,087	4,087	2.2024	9,001	34.00	5,941	100	0	5,941	0
40	Addition Metro Con	1975	50	1,421,442	1,421,442	2.2024	3,130,584	34.00	2,066,185	100	0	2,066,185	0
41	Line ext & renewal	1975	50	144,764	144,764	2.2024	318,828	34.00	210,426	100	0	210,426	0
42	Water line ext & ren	1976	50	158,417	158,417	2.0317	321,856	32.00	218,862	100	0	218,862	0
43	Routine line ext	1978	50	95,538	95,538	1.9019	181,704	28.00	130,827	100	0	130,827	0
44	Routine line ext	1978	50	243,088	243,088	1.7633	428,637	28.00	308,619	100	0	308,619	0
45	Cap const in prog	1978	50	1,302,472	1,302,472	1.7633	2,296,649	28.00	1,653,587	100	0	1,653,587	0
46	Capitalize C/P	1978	50	131,621	131,621	1.7633	232,087	28.00	167,103	100	0	167,103	0
47	Remod wksp main pl	1978	35	10,365	10,365	1.7633	18,277	40.00	10,966	0	0	0	0
48	Routine line ext	1979	50	132,682	132,682	1.6111	213,764	26.00	158,185	100	0	158,185	0
49	Completed const	1980	50	518,781	518,781	1.4985	777,393	24.00	590,819	100	0	590,819	0
50	Gordy cons Div/2	1981	50	525,583	525,583	1.3691	719,576	22.00	561,269	100	0	561,269	0
51	Am C Iron Div 1/2/3	1981	50	3,620,690	3,620,690	1.3691	4,957,087	22.00	3,886,528	100	0	3,886,528	0
52	Renewal ext aud/ent	1981	50	61,322	61,322	1.3691	83,956	22.00	65,486	100	0	65,486	0
53	Capeng Div I,II,III	1981	50	66,875	66,875	1.3691	91,559	22.00	71,416	100	0	71,416	0
54	Div1/Div3 Pew Cons	1981	50	1,011,494	1,011,494	1.3691	1,384,836	22.00	1,080,172	100	0	1,080,172	0
55	Capitalize improvements	1982	50	35,582	35,582	1.2808	45,573	20.00	36,458	100	0	36,458	0
56	N Cols HSPS	1982	35	833,282	833,282	1.2808	1,067,268	28.57	762,350	100	0	762,350	0
57	Cptiz improvements	1982	50	16,463	16,463	1.2808	21,086	20.00	16,869	100	0	16,869	0
58	Fence and install	1982	35	6,100	6,100	1.2808	7,813	28.57	5,581	100	0	5,581	0
59	Capitalize routine ln ext	1982	50	41,536	41,536	1.2808	53,199	20.00	42,559	100	0	42,559	0
60	N Col high pres pump	1982	35	51,845	51,845	1.2808	66,403	28.57	47,432	100	0	47,432	0
61	Pump repair	1982	35	15,206	15,206	1.2808	19,476	28.57	13,912	0	0	0	0
62	Capitalize routine ln ext	1982	50	136,152	136,152	1.2808	174,383	20.00	139,506	100	0	139,506	0

#	Description	Year												
63	Capitalize eng JJG	1983	50	35,371	35,371	1.2309	43,538	18.00	100	35,701	100	0	35,701	0
64	Dist Sys Aud 5-83	1983	50	60,286	60,286	1.2309	74,206	18.00	100	60,849	100	0	60,849	0
65	Audit 5 Cap wat ext	1983	50	154,797	154,797	1.2309	190,540	18.00	100	156,243	100	0	156,243	0
66	NE Cols 30 Main PW	1984	50	715,324	715,324	1.2208	873,268	18.00	100	733,545	100	0	733,545	0
67	N Cols By pass	1984	50	1,114,952	1,114,952	1.2208	1,361,133	16.00	100	1,143,352	100	0	1,143,352	0
68	Drill/Tap Mach	1984	35	1,527	1,527	1.2208	1,864	22.86	100	1,438	100	0	1,438	0
69	FP trans repairs	1984	35	13,973	13,973	1.2208	17,058	22.86	0	13,159	100	0	0	0
70	Cap Waste Ltd	1985	35	79,000	79,000	1.1965	94,523	20.00	100	75,618	100	0	75,618	0
71	Cap control flow	1985	35	9,000	9,000	1.1965	10,768	20.00	100	8,614	100	0	8,614	0
72	Dis sys audit 21-85	1985	50	74,403	74,403	1.1965	89,023	14.00	100	76,560	100	0	76,560	0
73	Dist audit 40-6/86	1986	50	198,369	198,369	1.1627	230,644	12.00	100	202,967	100	0	202,967	0
74	Ridgewood system	1987	50	13,376	13,376	1.1297	15,111	10.00	100	13,600	100	0	13,600	0
75	Shepard Dr pump st	1987	35	8,521	8,521	1.1297	9,626	14.29	100	8,250	100	0	8,250	0
76	Audit entry 21 A	1987	50	166,319	166,319	1.1297	187,891	10.00	100	169,102	100	0	169,102	0
77	400 AMP svc & wire	1987	35	6,823	6,823	1.1297	7,708	14.29	100	6,607	100	0	6,607	0
78	2 Pumps	1987	35	16,500	16,500	1.1297	18,640	14.29	100	15,976	100	0	15,976	0
79	Kelly Tamps & Hammer	1987	35	4,880	4,880	1.1297	5,513	14.29	100	4,725	100	0	4,725	0
80	Jenkins Rd	1988	50	33,572	33,572	1.1034	37,043	8.00	100	34,080	100	0	34,080	0
81	Upatoi Lane	1988	50	46,699	46,699	1.1034	51,528	8.00	100	47,406	100	0	47,406	0
82	NE Col Water-Gordy	1988	50	2,518,271	2,518,271	1.1034	2,778,660	8.00	100	2,556,367	100	0	2,556,367	0
83	Audit entry 18	1988	50	379,491	379,491	1.1034	418,730	8.00	100	385,232	100	0	385,232	0
84	Tanks	1989	50	8,227	8,227	1.0798	8,884	6.00	100	8,351	100	0	8,351	0
85	Filter plant Phs I	1989	50	9,681,130	9,681,130	1.0798	10,453,684	6.00	100	9,826,463	100	0	9,826,463	0
86	AJE #9 misc projects	1989	35	1,063,760	1,063,760	1.0798	1,148,648	8.57	100	1,050,209	100	0	1,050,209	0
87	JJG eng fee	1989	50	6,376	6,376	1.0798	6,885	6.00	100	6,472	100	0	6,472	0
88	Cap routine line X	1990	50	320,708	320,708	1.0590	339,630	4.00	100	326,045	100	0	326,045	0
89	AJE #8 routine line	1990	50	218,452	218,452	1.0590	231,341	4.00	100	222,087	100	0	222,087	0
90	McKee Rd	1990	50	9,556	9,556	1.0590	10,120	4.00	100	9,715	100	0	9,715	0
91	Russell Rd	1990	50	21,170	21,170	1.0590	22,419	4.00	100	21,522	100	0	21,522	0
92	Water plant imp II	1991	50	9,536,709	9,536,709	1.0348	9,868,586	2.00	100	9,671,214	100	75	2,417,804	7,253,411
93	Chattsworth Road	1991	50	5,327	5,327	1.0348	5,512	2.00	100	5,402	100	100	0	5,402
94	9-1 pwr dr v/v opr	1991	35	3,382	3,382	1.0348	3,500	2.86	100	3,400	100	100	0	3,400
95	Fall Line Freeway	1991	50	193,456	193,456	1.0348	200,188	2.00	100	196,184	100	100	0	196,184
96	WTR Boostr & pow P	1991	35	11,164	11,164	1.0348	11,553	2.86	100	11,223	100	100	0	11,223
97	Routine Line Ext	1991	50	264,681	264,681	1.0348	273,892	2.00	100	268,414	100	100	0	268,414
98	High Service P.S.	1991	35	2,881,684	2,881,684	1.0348	2,981,967	2.86	100	2,896,683	100	100	0	2,896,683
99	8 Watermain	1991	50	9,864	9,864	1.0348	10,207	2.00	100	10,003	100	100	0	10,003

Table 2B Inventory and Current Value of Water Facilities (Continued)

No.	Facility Name/location	Year built	Useful life (yrs)	Total amount paid ($)	Amount locally paid ($)	ENR factor 1992	Local replacement value 1992 ($)	Deprec percent 1992 (%)	Current local asset value 1992 ($)	Percent system improvement (%)	Percent growth related (%)	Current value local systemwide improvements ($)	Current value local growth related improvements ($)
100	So Col feeder main	1991	50	4,725,000	4,725,000	1.0000	4,725,000	2.00	4,630,500	100	100	0	4,630,500
101	River PS expansion	1991	35	950,000	950,000	1.0000	950,000	2.86	922,830	100	100	0	922,830
102	SFDA studies	1991	50	300,000	300,000	1.0000	300,000	2.00	294,000	100	100	0	294,000
103	No Col PS expansion	1991	50	300,000	300,000	1.0000	300,000	2.00	294,000	100	100	0	294,000
104	Wtr main ext Ph I	1991	50	360,000	360,000	1.0000	360,000	2.00	352,800	100	100	0	352,800
105	Ind Park water lines	1991	50	630,000	630,000	1.0000	630,000	2.00	617,400	100	100	0	617,400
106	Secondary pwr sup	1992	35	28,081	28,081	1.0000	28,081	0.00	28,081	100	100	0	28,081
107	1975 Bond issue costs	1975	20	31,000	31,000	2.2024	68,274	85.00	10,241	100	0	10,241	0
108	1990 Bond issue costs	1990	20	54,000	54,000	1.0590	57,186	10.00	51,467	100	100	0	51,467
	Total system value			62,933,308	62,933,308		150,240,672		76,929,393			59,055,559	17,835,798

Notice that in earlier years the level of system expansion detail is less than in later years. This is common among many systems, including some of the nation's largest systems. However, in the examples given, records were available in enough detail to know that the improvements listed in early years were indeed related to system expansions with relatively long useful lives. For any item, the inventory need not extend further back in time than its useful life, even if it is still in use.

The inventory must be done by service area. The inventory need not include components inside subdivisions or developments that are designed to serve only those projects, since they are installed by developers. The inventory could include these items, but would need to show them not as system improvements (see below). However, where the system replaces those lines using resources from ratepayers, those lines may be relisted in the inventory as new system improvements. The argument is that, for the most part, collection and transmission systems do not work efficiently if parts of the system fail. Moreover, failure in part of the system may inefficiently shift development burdens demands to other parts of the system.

Notice also that there are four major categories of items. The first are the system improvements themselves, including such items as treatment plants, pumps, lines, expansions, replacements, and so forth. These items compose the bulk of the listings; but there are three other categories. One includes studies. Although small in overall cost relative to the physical elements of the system, the efficiency of the entire system depends on them. Studies are thus included as inventory items. Also included are the costs associated with issuing debt. Debt instruments help soften the burden of paying for the lumpy nature of water and wastewater investments, but the processing of those instruments is not free. Those costs are legitimately inventoried and assessed for facility expansion fee calculation purposes. Finally, there are other system assets that may not be accounted for in the inventory, such as buildings, land and yard areas, vehicles, and equipment. The replacement costs of these assets (less depreciation) are usually reported in the insurance policies for the current year. A single line can be added to the inventory to reflect these additional items and their associated values in appropriate columns.

Year Built

The year in which each item's cost was incurred is shown. If an item is replacing another, the newer item is shown and the original item is removed. This column will be used later to help determine current system value after accounting for depreciation.

Useful Life (Years)

The expected useful life of each component is shown here. Useful life is given by various manufacturer's reports, studies, and local knowledge. The useful life of studies is given 50 years in this example inasmuch as those studies typically plan for 50-year useful lives of major system components. They are given maximum, reasonable, useful lives on the assumption that most studies result in system improvements that will benefit the system at least that long. The useful lives of debt issuance costs are set equal to the debt service period because those costs are amortized during debt service, but not beyond. (Even if the instrument is refinanced later, these costs remain listed and associated with the original term of the instrument.) This column will be used later to help determine current system value after accounting for depreciation.

Total Amount Paid

This is the total amount paid by all sources of revenue at the time of installation of each listed item, including federal grants, state grants, local revenues, developer-installed improvements, and sources and revenues generated through sale of bonds. This column satisfies *rational nexus* concerns about determining the cost of existing facilities.

Amount Locally Paid

This is the amount paid by local government and thus excludes grants from federal and state sources and improvements installed by developers. The amounts shown are the actual costs incurred by local ratepayers or taxpayers. This column satisfies *rational nexus* concerns about determining how existing facilities were financed in the past.

Engineering News & Record (ENR) Value Factor

A crucial part of satisfying *rational nexus* criteria is adjusting for the time-value of money. For facilities installed in the past, there may be several ways in which to arrive at a current value. One way is to undertake new engineering cost estimates for each inventory item, then adjust for depreciation. But engaging in new engineering cost estimates every year for all existing parts of the system can be cumbersome and expensive. Moreover, as a practical matter, some facilities could not be rebuilt with the same materials or specifications involved in the original installation and, thus, cost estimates would be difficult to determine.

Although annual engineering cost estimates for each inventory item may be preferable, it should not be necessary to be consistent with the *reasonableness* requirements of the *rational nexus.* Instead, adjustments to current value may be made using price indices accounting for inflation. The most common are the price indices published monthly by the Bureau of Labor Statistics under various consumption and production considerations.

However, the price index selected for this analysis is the 20-Cities Construction Cost Index calculated monthly since 1917 by *Engineering News and Record.* The ENR Index is a general construction cost index based on a hypothetical unit of construction requiring: 6 barrels of portland cement, 1088 million board ft of 2" × 4" lumber, 2500 lb of structural steel, 68.38 h of skilled labor, and 200 h of common labor. The ENR Construction Cost Index Factor represents the number of fourth quarter 1992 construction units that could be purchased with the cost in that year of one construction unit, *Engineering News-Record,* monthly. The quarterly ENR index figures and factors in 1992 dollars are reported in Appendix 1. (These figures would be adjusted annually. The quarter selected is usually based on the fiscal year used.)

Local Replacement Value

This column establishes the current replacement cost of each inventory item based only on the local share of the cost incurred and assuming that the item may be reasonably replaced at a price equal to the original cost times the ENR factor. For each item, the replacement value in terms of the local share of cost is calculated as:

(Original cost, Local share) × (ENR factor)

Depreciation Percent

This column shows the proportion, in percent, of the inventory item that has depreciated in straight-line fashion from the original installation year to the current year. For each item, the percent depreciation is calculated as:

$$\frac{(Current\ year) - (Installation\ year)}{Useful\ life\ years}$$

Current Local Asset Value

This column shows the current value of each inventory item in terms of the local share of the original cost. The column demonstrates consistency of analysis with the *rational nexus* criteria because it adjusts the value of the local share of system costs considering the time-value of money and the wear-and-tear incurred since installation. For each item, the current value is calculated as:

(Replacement value, Local share) × (1 − Depreciation percent)

Percent System Improvement

This column shows the percent of the item that serves the service area as a whole and is not dedicated for use by a particular development. Qualifying items may extend service to the boundary of a particular development, such as an industrial park, but do not usually include service within such developments. However, if lines within the industrial park become the responsibility of the system to maintain and replace, then the item is considered a system improvement.

Almost every inventory item is considered 100% system improvement. Notice, however, that some items are shown with varying percentages of system improvement status. Where the developer installs pieces of the system, or grants are used to finance pieces of the system, only the local share of that cost is considered. In some situations, particular development projects and governmental agencies will agree to share in the cost of installing, maintaining, and replacing certain improvements.

Percent Growth Related

This column reports the percent of the item that is reasonably apportioned to accommodate development in the future. This is customarily determined for the initial year of analysis for facility expansion fee calculation purposes, in this case, 1991. It indicates investments made in the relatively recent past specifically to accommodate new development. After a period of time, the percent growth related assessment should be reduced toward zero, and one day become zero when the growth being served by this item has occurred.

Current Value, Local Systemwide Improvements

This column shows for each inventory item its current value considering its status as a system improvement and considering the share of the costs incurred by the local government. For each inventory item, it is calculated as:

$$[(\text{Current value, Local share}) \times (\text{Percent system improvement}) \\ \times (1 - \text{Percent growth related})]$$

Current Value, Local Growth-Related Improvements

This column shows for each inventory item its current value considering its status as a system improvement installed to serve new development and considering the share of the costs incurred by the local government. For each inventory item, it is calculated as:

$$[(\text{Current value, local share}) \times (\text{Percent system improvement}) \\ \times (\text{Percent growth related})]$$

ASSESSMENT OF EXISTING FACILITIES

From information supplied in Tables 1 and 2A,B, the assessment of the current system capacities to accommodate future development can be made. In addition, this information can then be used to help calculate the current value of the system per unit of capacity. Table 3 makes this assessment for wastewater and water facilities. In this example, the table shows that the existing system has some excess capacity to accommodate new development and that, in the case of

Table 3 Assessment of Existing and Planned Wastewater and Water Capacity

Assessment consideration	Wastewater	Water
System capacity, 1991, gallons	40,000,000	55,000,000
Growth-related expansions, 1991–1999, gallons	8,000,000	0
Total planned capacity, gallons	48,000,000	55,000,000
Average daily demand, gallons	24,700,000	29,900,000
Excess capacity, 1991, gallons	15,300,000	25,100,000

wastewater facilities, expansions are planned in the near future to accommodate more development.

Note is made of the selection of years for which data are reported. The City has determined that because it made major investments in water and wastewater systems in 1991 to accommodate new development, this is the appropriate base year in which to calculate the value of excess capacity as shown in Table 3. With each passing year, "excess" capacity diminishes and, as a result, system development charges need to be revised. However, the City has decided to use excess capacity in 1991 as the base year on which to base the value of such capacity dedicated to accommodating new development. At some point in the future, probably when average daily use has absorbed all excess capacity as of 1991, a new base year may be established for the calculation of SDCs.

FACILITY EXPANSION COSTS

The next step is an inventory of wastewater and water facilities planned for installation according to the capital improvements program during the forthcoming 6 years, although longer periods are desired. This inventory is prepared for each service area; but since Columbia has one service area, only one set of tables is prepared for each system. Tables 4A and 4B present the relevant data and analysis. The meaning of each column is described here.

Facility Name/Location

This is the list of planned additions to the existing system. Where there is more than one service area, the inventory must be done by service area. The inventory need not include components inside

Table 4A Wastewater Capital Improvements Program

CIP#	Projection description	Year to be built	Useful life (yrs)	Total cost ($)	Local cost ($)	Percent system improvement (%)	Percent growth related (%)	Local costs, system-wide improvements ($)	Local costs, growth-related improvements ($)
WWT2	Land application site purchase	1994		5,000,000	5,000,000	100	30	3,500,000	1,500,000
IES5	Cooper Branch sewer extension	1995	50	690,000	690,000	100	100	0	690,000
IES6	Cooper Creek sewer extension	1995	50	200,000	200,000	100	100	0	200,000
IES8	Stevens Lane sewer	1995	50	105,000	105,000	100	100	0	105,000
CS4	South Lumpkin Road sewer	1995	50	1,400,000	1,400,000	100	100	0	1,400,000
CS9	Woodruff Farm Road sewer	1996	50	950,000	950,000	100	100	0	950,000
IES9	Hunter/Schatulga Road sewers	1996	50	100,000	100,000	100	100	0	100,000
CS6	Parallel trunk sewer-Moss Drive	1996	50	1,040,000	1,040,000	100	100	0	1,040,000
CS5	Standing Boy Creek sewer (eastside)	1996	50	450,000	450,000	100	100	0	450,000
CS5	Standing Boy Creek sewer (westside)	1997	50	530,000	530,000	100	100	0	530,000
CS3	Industrial Park sewer lines	1997	50	3,300,000	3,300,000	50	0	1,650,000	0
CS2	Heiferhorn Creek sewer extension	1997	50	3,325,000	3,325,000	100	100	0	3,325,000
WWT2	Sewer line extensions	1997	50	10,000,000	10,000,000	100	100	0	10,000,000
WWT2	Panhandle lines, pumps, force mains	1998	50	20,000,000	20,000,000	100	100	0	20,000,000
WWT2	Parallel sewer lines	1998	50	15,000,000	15,000,000	100	100	0	15,000,000
WWT2	Parallel force mains	1998	50	5,000,000	5,000,000	100	100	0	5,000,000
WWT2	Expand pump station capacities	1998	50	4,000,000	4,000,000	100	100	0	4,000,000
WWT3	Expand WWTF to 48 MGD	1999	50	12,000,000	12,000,000	100	100	0	12,000,000
WWT3	Expand sludge facilities	1999	50	3,000,000	3,000,000	100	100	0	3,000,000
IES16	Provide phosphorus removal	1999	50	14,125,000	14,125,000	100	0	14,125,000	0
IES17	Provide ammonia removal	1999	50	6,000,000	6,000,000	100	0	6,000,000	0
	1985 Bond remaining interest, PV'93				17,798,909	100	0	17,798,909	0
	1991 Bond remaining interest, PV'93				10,716,396	100	71	3,107,755	7,608,641
	1993 Bond issue interest, PV'93				62,064,493	100	76	14,895,478	47,169,015
	Total wastewater facilities			106,215,000	196,794,798			61,077,142	134,067,656

Table 4B Wastewater Capital Improvements Program

CIP#	Projection description	Year to be built	Useful life (yrs)	Total cost ($)	Local share of total cost ($)	Percent system improvement (%)	Percent growth related (%)	Local costs, system-wide improvements ($)	Local costs, growth-related improvements ($)
DS7	East Columbus feeder main and tank site	1993	50	1,050,000	1,050,000	100	100	0	1,050,000
WT5	Upgrade PAC feed system	1993	50	300,000	300,000	100	25	225,000	75,000
IES1	Filter press plates	1994	50	500,000	500,000	100	100	0	500,000
DS9	Water main extension	1994	50	752,000	752,000	100	100	0	752,000
DS8	East Columbus storage tank	1995	50	1,235,000	1,235,000	100	100	0	1,235,000
DS3	Industrial Park water mains	1996	50	1,361,000	1,361,000	50	0	680,500	0
DS4	Water storage tanks in gravity system	1997	50	5,000,000	5,000,000	100	0	0	5,000,000
	Baffle clear wells at filter plant	1998	50	500,000	500,000	100	0	500,000	0
	Extend lines to NE, add tank	1998	50	8,000,000	8,000,000	100	100	0	8,000,000
	Extend water lines	1999	50	4,000,000	4,000,000	100	100	0	4,000,000
	New water storage tanks	1999	50	1,000,000	1,000,000	100	100	0	1,000,000
	1990 Bond remaining interest, PV'93				10,514,890	100	42	6,098,636	4,416,254
	1993 Bond interest, PV'93				13,847,426	100	94	830,846	13,016,580
	Total water facilities			23,698,000	23,698,000			8,334,982	39,044,834

subdivisions or developments that are designed to serve only those projects inasmuch as they are installed by developers. There may be three major categories of items: (1) system improvements, including such items as treatment plants, pumps, lines, expansions, and replacements; (2) studies planned to be undertaken; and (3) the cost of issuing future debt service instruments, which can only be a reasonable estimate since they have yet to be incurred.

It is important to update the CIP annually for one simple reason. When a project is completed, it is no longer part of the CIP in terms of planned improvements. The item must be transferred to the inventory of existing facilities and listed with relevant information such as actual total costs, the local share of those costs, useful life in years, and percent of the item considered a system improvement.

Year To Be Built

This column shows the year of projected installation. In the case of projects extending over several years, the beginning or completion years need only be shown for purposes of calculating SDCs. It is customary to report the beginning year because this is the year in which commitments are made to install the facility.

Useful Life in Years

This column is the same as for the inventory of existing facilities.

Total Amount To Be Paid

This column shows the amount expected to be paid by all sources of revenue at the time of installation. This is only an estimate.

Amount To Be Locally Paid

This column shows the amount expected to be paid by local sources, not including private donations, contributions, and exactions, but including general fund, enterprise fund, and debt financing even if the loan comes from federal or state sources.

Percent System Improvement

This column shows the percent of the item that serves the service area and is not dedicated for use by a particular development. Because of considerations raised in the inventory of existing facilities, almost every inventory item is considered 100% system improvement. (Remember that where the developer installs pieces of the system, or grants are used to finance pieces of the system, only the local share of that cost is considered.) Some items may have varying percentages of system improvement status. In these situations, particular development projects and governmental agencies may have agreed to share in the cost of installing, maintaining, and replacing certain improvements.

Percent Growth Related

This column shows the percent of the item that is intended to accommodate growth through expansion of system capacity within the service area.

Local Systemwide Improvement Costs

An important point must be made here about CIP items. Once installed, a CIP item must be moved from the CIP to the inventory of existing systems. This should be done on an annual basis, concurrent with the annual updates to the CIP. However, once it is moved from the CIP to the inventory, the nature of its relationship to the system changes. Some items in the CIP may not be related to accommodating growth, but are system improvements. In this example, a number of projects listed in Columbia's CIP cannot be totally financed from system development charges. However, because they are considered system improvements, they will be added to the current value of the existing system. This additional value must be accounted for in the calculation of SDCs. For each CIP item not fully accounted for as a growth-related cost, the system improvement related costs are calculated as:

[(Amount to be locally paid) × (Percent system improvement)]
– (Growth-related costs)

The figure will always be equal to or greater than $0.

Local Growth-Related Costs

This column shows the amount of the improvement that may be eligible for financing in part by expansion-related SDCs. For each CIP item, it is calculated as:

(Amount to be locally paid) × (Percent growth related)

Note must be made that Tables 4A and 4B include interest costs for outstanding and planned debt — in this case, bond issues. These figures come from Tables 7A, 7B, and 7C for outstanding debt, and Tables 8A and 8B for anticipated future debt to help finance the CIP. The interest used in the CIP is the present value of the remaining interest payments to be made on outstanding debt, and the present value of anticipated interest payments using the discount rate shown on the debt service tables. The reason capital payments are not included is that in the case of outstanding debt, capital payments are reflected in the inventory of existing facilities; and in the case of anticipated debt, the payments are reflected in the facilities to be financed from the CIP.

GROSS SYSTEM DEVELOPMENT CHARGES

At this point, it is possible to calculate gross system development costs attributable to new development. This is shown in Table 5. As new development is added to the wastewater and water systems, it enjoys the use of the investments previously incurred by existing ratepayers and taxpayers. An equitable SDC recognizes this and includes a component reflecting the proportionate share of the current asset value of the system that may be recouped from new development. A second component of the SDC reflects the growth-related investments being made explicitly to accommodate new development.

The first component involves determining the appropriate charge per gallon for "systemwide" improvements planned or in place. It is usually the situation that when systems expand, new development is served by facilities built and paid for by existing development. In the case of Columbia, new development is substantially found at the fringe. Wastewater generated by new development must be transmitted from the urban fringe through facilities built and paid for by existing development. Thus, new development uses collection facili-

Table 5A Gross Wastewater System Development Charge per Gallon

Assessment consideration	Wastewater
System capacity, 1991, gallons	40,000,000
Planned system capacity, gallons	48,000,000
Current local asset value, systemwide improvements	$93,966,655
Local CIP systemwide improvements	$61,077,142
Existing & planned systemwide improvements	$155,043,797
Existing & planned systemwide improvements per gallon (48,000,000 capacity)	$3.23
Planned growth-related expansion, gallons	8,000,000
Excess capacity, 1991, gallons	15,300,000
Total excess and planned growth-related capacity, 1991, gallons	23,300,000
Current asset value, local growth-related improvements	$11,971,730
CIP growth-related costs	$134,067,656
Total growth-related costs	$146,039,386
Growth-related expansion cost/gallon, (23,300,00 capacity)	$6.27
Gross wastewater system development charge per gallon	$9.50

Table 5B Gross Water Facility System Development Charge per Gallon

Assessment consideration	Water
Existing system capacity, gallons	55,000,000
Planned system capacity, gallons	55,000,000
Current local asset value, not growth related	$59,055,559
Local CIP system improvements, not growth related	$8,334,982
Existing & planned improvements, not growth related	$67,390,541
Existing & planned cost/gallon, not growth related	$1.23
Planned growth-related expansion, gallons	0
Excess capacity, 1991, gallons	25,100,000

Table 5B Gross Water Facility System Development
Charge per Gallon (Continued)

Total excess and planned growth-related capacity, gallons	25,100,000
Existing growth-related recoupment value	$17,835,798
Local CIP growth-related costs	$39,044,834
Total growth related costs	$56,880,632
Growth-related expansion cost/gallon, (25,100,000 capacity)	$2.27
Gross water system development charge per gallon	$3.50

ties installed at the urban fringe *and* facilities found in developed areas. On the other hand, some infill and redevelopment within developed areas also occurs. Such development often requires new or larger collection lines and related facilities, and also generates wastewater that passes through other facilities built and paid for by existing development. The same conditions hold for water facilities. Such facilities are called "systemwide" facilities because they serve all development, *both* existing and new. A new customer entering the system enjoys the benefits of previously incurred and planned investments to provide wastewater and water service throughout the service area. Unless new development is charged for this benefit, new customers become "free-riders". For these reasons, it is appropriate to consider that new development must be held financially accountable for its proportionate share of the cost of systemwide facilities. The CIP also shows planned expansions or improvements of a systemwide nature, such as new or larger mains, upgraded treatment, and expanded facilities.

The first SDC component is the sum of the current asset value of systemwide improvements financed from local resources ("current local asset value, systemwide improvements" in Table 5) and planned systemwide improvements financed locally as shown in the CIP ("local CIP systemwide improvements" in Table 5). This total is divided by total planned system capacity at the last year of the CIP, which is 1999 in this example. The result is the proportionate share of the current value of existing and planned investments for systemwide improvements apportioned to new development.

The second component involves attributing existing and new facilities that are primarily used by new development. This would include all expansions to treatment capacity and related facilities. It also includes facilities needed to make excess capacity accessible to new development; otherwise, excess capacity could not be used."Growth-related" improvements include those installed since 1991, the base year for SDC calculation selected by Columbia, and planned as shown in the CIP. The second SDC component is thus the sum of the current asset value of growth-related improvements financed from local resources ("current local asset value, growth-related improvements" in Table 5) and planned growth-related improvements financed locally as shown in the CIP ("local CIP growth-related improvements" in Table 5). This total is divided by excess and planned new system capacity. The result is the proportionate share of the current value of and planned investments for growth-related improvements attributable to new development.

DEBT SERVICE COSTS AND CREDITS

To this point, only gross SDCs have been calculated on a proportionate share basis. The *rational nexus* criteria require consideration of credits because new development cannot be charged twice for system improvements required to accommodate its impacts — once with system development charges and then again through tax revenues or user fees applied to debt. Columbia Water Works finances capital improvements with revenue bonds. Revenue bonds are retired with water and wastewater rate revenues. Since new customers will be helping to retire outstanding bonds that were used to create existing capacity, their SDCs should be reduced by the present value of the share of their future rates that will be used to retire outstanding bonds.

(Sometimes, local governments issue general obligation bonds that are retired through property tax payments, some payments of which come from the vacant land on which new development has yet to be built and connected to the system. General obligation bond debt service thus adds the dimension of giving credit for both *past* and future payments made through property taxes. An example of how costs and credit for general obligations bonds is calculated in shown in Appendix 2.)

Columbia Water Works currently has three outstanding bond issues, including two revenue bonds for wastewater facilities and one revenue bond for facilities. The outstanding debt is summarized in Table 6.

Table 6 Outstanding Water/Wastewater Debt

Debt instrument	Outstanding principal	
	Water projects	**Wastewater projects**
1985 Series revenue bonds		$53,951,042
1990 Series revenue bonds	$21,049,767	
1991 Series revenue bonds		$20,138,500
Total	$21,049,767	$74,089,542

In addition to the outstanding debt, credit should also be given for the present value of future debt payments for future bonds that is anticipated to be used to construct planned capital improvements.

However, there are costs associated with debt service that must be accounted. When debt is incurred, the local government receives funds it then uses to invest in systems. Thus, the entire principal amount of the debt can be viewed as being capitalized into existing facilities. The inventory and assessment sections of the SDC analysis determines the current value of existing facilities. However, future interest payments must be made by all existing and future ratepayers or taxpayers. Interest costs must therefore be counted as a cost to be borne by all users. Interest costs for outstanding and anticipated debt have been included in Table 4. In all cases, the present value of the stream of all future interest payments is calculated and reported in Table 7 for outstanding debt and in Table 8 for future anticipated debt, and summarized in Table 4.

The interesting feature of present value analysis is that the lower the discount rate, the higher the present value of interest costs and thus the higher the potential SDC if interest is considered a cost yet to be borne by ratepayers; but then also, the higher the credit for total debt service payments to be made by new development. Conversely, the higher the discount rate, the lower the present value of interest costs and total debt service credits. In the example case, a discount rate of 6% is used because it is the tax exempt bond yield rate for the City of Columbia at the time the SDCs were calculated. Developers may argue for lower discounts rate such as passbook savings rates or the most recently published inflation rates (the consumer price index published monthly by the U.S. Bureau of Labor Statistics in *Monthly Labor Review*). On the other hand, local governments may counter that the appropriate discount rate is the private sector's true cost of money, which is often the prime lending rate plus two or three points. The discount rate that seems to be fairest is the local government's tax exempt borrowing rate for long-term debt or the rate it pays on tax anticipation notes.

Tables 7A, B, and C show the outstanding revenue bonds for wastewater and water facilities. Where bonds are issued to pay for both wastewater and water facilities, they must be apportioned between those systems. Where bonds are issued covering multiple service areas, they must be apportioned among the service areas inasmuch as system development charges must be calculated for each service area. The meaning of each column is described below.

Year. This column shows the year of debt service payments. In cases where debt service is in semiannual or quarterly payments, this column may be subdivided into more periods.

Principal, Interest, Total Payments. These columns come directly from the bond perspectus for the debt service.

Consumption (MGD). This column reports the actual or projected consumption as measured at the treatment plants; in this case, on an average daily basis based on annual figures.

Debt Service Per Gallon. This column shows the annual average debt service per gallon treated on an average daily basis during the year. It is computed as:

$$\text{(Total debt service payment)} / \text{(Consumption in MGD} \times 1{,}000{,}000)$$

The net present value of debt service payments and payments per gallon is shown at the bottom of the tables. The interest costs must be borne by all ratepayers and have not been accounted for in either the inventory of existing facilities (which capitalizes the principal received from the sale of bonds) or the CIP. New development thus must be apportioned its proportionate share of the present value of future interest payments. However, new development is also credited with the present value of all future principal and interest debt service payments.

The next step is estimating debt service costs and credits for debt to be issued to finance the capital improvements program. Since no debt service schedules are usually prepared for future bond issues, assumptions must be developed to estimate an appropriate credit. Typical assumptions are that (1) all capital improve-

Table 7A 1985 Revenue Bond Series for Wastewater

Year	Principal payment	Interest payment	Total payment	Daily use (MGD)	Debt service per gallon
1985	$890,768	$2,986,748	$3,877,516	24.3	$0.1596
1986	$1,332,803	$3,546,482	$4,879,285	25.5	$0.1913
1987	$1,399,778	$3,481,870	$4,881,648	26.6	$0.1835
1988	$1,473,450	$3,409,887	$4,883,337	27.1	$0.1802
1989	$1,553,820	$3,330,771	$4,884,591	28.2	$0.1732
1990	$1,640,888	$3,238,104	$4,878,991	29.0	$0.1682
1991	$1,738,671	$3,138,002	$4,876,673	24.7	$0.1974
1992	$1,844,492	$3,029,628	$4,874,120	27.2	$0.1792
1993	$1,959,689	$2,912,222	$4,871,911	28.2	$0.1728
1994	$2,088,281	$2,787,396	$4,875,676	29.0	$0.1681
1995	$2,223,570	$2,653,806	$4,877,376	29.8	$0.1637
1996	$2,372,255	$2,508,524	$4,880,778	30.6	$0.1595
1997	$1,730,634	$2,354,232	$4,084,866	31.5	$0.1297
1998	$2,709,809	$2,187,781	$4,897,590	32.4	$0.1512
1999	$2,902,697	$1,996,732	$4,899,428	33.3	$0.1471
2000	$3,111,659	$1,791,959	$4,903,617	34.3	$0.1430
2001	$3,336,695	$1,572,288	$4,908,982	35.3	$0.1391
2002	$4,783,355	$1,335,594	$6,118,949	36.3	$0.1686
2004	$6,243,410	$613,191	$6,856,600	38.4	$0.1786
2005	$2,774,105	$188,639	$2,962,744	39.5	$0.0750
Total	$53,951,042	$50,074,181	$104,025,222		$2.52
NPV @ 6% discount					
1993	$26,665,547	$17,798,909	$44,464,456		$1.36

ments listed in the CIP will be financed with long-term debt; (2) all debt instruments will be issued in the first year of the CIP; and (3) the debt service schedule will be similar to the schedule for the most recently issued bond series or that of a comparable local government that recently issued revenue bonds for similar purposes. While somewhat simplistic, these assumptions are necessary to reasonably address credit issues. These assumptions will probably lead to a slightly higher credit than may actually be due

Table 7B 1991 Revenue Bond Series for Wastewater

Year	Principal payment	Interest payment	Total payment	Daily use (MGD)	Debt service per gallon
1991	$332,500	$1,114,874	$1,447,374	24.7	$0.0586
1992	$497,500	$1,323,808	$1,821,308	27.2	$0.0670
1993	$522,500	$1,299,690	$1,822,190	28.2	$0.0646
1994	$550,000	$1,272,821	$1,822,821	29.0	$0.0629
1995	$580,000	$1,243,289	$1,823,289	29.8	$0.0612
1996	$612,500	$1,208,699	$1,821,199	30.6	$0.0595
1997	$649,000	$1,171,333	$1,820,333	31.5	$0.0578
1998	$688,500	$1,130,880	$1,819,380	32.4	$0.0562
1999	$731,500	$1,087,056	$1,818,556	33.3	$0.0546
2000	$779,500	$1,040,461	$1,819,961	34.3	$0.0531
2001	$830,000	$990,596	$1,820,596	35.3	$0.0516
2002	$885,500	$936,366	$1,821,866	36.3	$0.0502
2003	$646,000	$878,773	$1,524,773	37.3	$0.0409
2004	$1,011,500	$816,641	$1,828,141	38.4	$0.0476
2005	$1,083,500	$745,327	$1,828,827	39.5	$0.0463
2006	$1,161,500	$668,891	$1,830,391	40.6	$0.0451
2007	$1,245,500	$586,894	$1,832,394	41.8	$0.0438
2008	$1,785,500	$498,542	$2,284,042	43.0	$0.0531
2009	$2,180,000	$377,128	$2,557,128	44.2	$0.0579
2010	$2,330,500	$228,888	$2,559,388	45.5	$0.0563
2011	$1,035,500	$70,414	$1,105,914	46.8	$0.0236
Total	$20,138,500	$18,691,370	$38,829,870		$0.80
NPV @ 6% discount					
1993	$9,941,751	$10,716,396	$20,658,147		$0.60

because all CIP projects are assumed to be financed from new debt, even though many systems will use capital reserve funds to pay for some improvements. If the local government does not anticipate issuing revenue bonds for capital expansion, or it has estimated the amount of bonds needed, it should take these determinations into consideration when calculating the costs and credits attributable to future debt.

**Table 7C 1990 Series Water Revenue Bond
Refinancing and Expansion**

Year	Principal payment	Interest payment	Total payment	Daily use (MGD)	Debt service per gallon
1990	$86,887	$291,331	$378,218	30.5	$0.0124
1991	$520,012	$1,383,711	$1,903,723	29.9	$0.0637
1992	$546,143	$1,358,501	$1,904,644	32.7	$0.0582
1993	$574,888	$1,330,416	$1,905,304	35.3	$0.0540
1994	$606,245	$1,299,548	$1,905,793	36.3	$0.0525
1995	$640,216	$1,263,392	$1,903,608	37.3	$0.0510
1996	$678,367	$1,224,336	$1,902,703	38.3	$0.0497
1997	$719,655	$1,182,053	$1,901,708	39.4	$0.0483
1998	$764,600	$1,136,245	$1,900,845	40.5	$0.0469
1999	$814,772	$1,087,542	$1,902,314	41.6	$0.0457
2000	$867,558	$1,035,420	$1,902,978	42.9	$0.0444
2001	$925,569	$978,736	$1,904,305	44.1	$0.0432
2002	$675,232	$918,537	$1,593,769	45.4	$0.0351
2003	$1,057,270	$853,594	$1,910,864	46.6	$0.0410
2004	$1,132,528	$779,053	$1,911,581	48.0	$0.0398
2005	$1,214,058	$699,158	$1,913,216	49.4	$0.0387
2006	$1,301,859	$613,450	$1,915,309	50.8	$0.0377
2007	$1,866,294	$521,101	$2,387,395	52.3	$0.0456
2008	$2,278,645	$394,193	$2,672,838	53.8	$0.0497
2009	$2,435,955	$239,245	$2,675,200	55.3	$0.0484
2010	$1,082,356	$73,600	$1,155,956	56.9	$0.0203
Total	$20,789,109	$18,663,162	$39,452,271		$0.69
NPV @ 6% discount					
1993	$10,468,970	$10,514,890	$20,983,860		$0.49

Tables 8A and B show the projected revenue bond costs and credits attributable to financing the CIP for wastewater and water facilities, respectively. If more than one service area is used, projections need to be made for each service area. The debt service schedules are created based on taking a straight proportionality to Columbia's most recently issued debt. This of course assumes that all

Table 8A Estimated Future Wastewater Revenue Bond

Year	Principal payment	Interest payment	Total payment	Daily use (MGD)	Debt service per gallon
1993	$1,753,680	$5,880,099	$7,633,779	28.2	$0.2707
1994	$2,623,927	$6,982,064	$9,605,991	29.0	$0.3312
1995	$2,755,783	$6,854,860	$9,610,643	29.8	$0.3225
1996	$2,900,824	$6,713,144	$9,613,969	30.6	$0.3142
1997	$3,059,051	$6,557,386	$9,616,437	31.5	$0.3053
1998	$3,230,463	$6,374,950	$9,605,414	32.4	$0.2965
1999	$3,422,973	$6,177,876	$9,600,849	33.3	$0.2883
2000	$3,631,305	$5,964,518	$9,595,823	34.3	$0.2798
2001	$3,858,096	$5,733,378	$9,591,474	35.3	$0.2717
2002	$4,111,259	$5,487,628	$9,598,887	36.3	$0.2644
2003	$4,377,608	$5,224,626	$9,602,233	37.3	$0.2574
2004	$4,670,327	$4,938,605	$9,608,932	38.4	$0.2502
2005	$3,407,150	$4,634,845	$8,041,995	39.5	$0.2036
2006	$5,334,880	$4,307,149	$9,642,029	40.6	$0.2375
2007	$5,714,624	$3,931,024	$9,645,648	41.8	$0.2308
2008	$6,126,013	$3,527,882	$9,653,896	43.0	$0.2245
2009	$6,569,048	$3,095,409	$9,664,457	44.2	$0.2187
2010	$9,417,130	$2,629,423	$12,046,554	45.5	$0.2648
2011	$11,497,813	$1,989,058	$13,486,871	46.8	$0.2882
2012	$12,291,584	$1,207,207	$13,498,791	48.0	$0.2812
2013	$5,461,461	$371,379	$5,832,840	48.9	$0.1193
Total	$106,215,000	$98,582,511	$204,797,511		$5.52
NPV @ 6% discount					
1993	$50,656,703	$62,064,493	$112,721,196		$3.22

cost characteristics of those revenue bonds apply to the next series to be issued.

The future costs and credits of those revenue bonds must be adjusted to properly attribute costs to new development. Recall that some CIP improvements are attributable mostly to new development, and new development must be attributed to the interest costs associated with growth-related improvements. However, new development

Table 8B Estimated Future Water Revenue Bond

Year	Principal payment	Interest payment	Total payment	Daily use (MGD)	Debt service per gallon
1993	$391,270	$1,311,929	$1,703,199	35.3	$0.0482
1994	$585,434	$1,557,793	$2,143,227	36.3	$0.0590
1995	$614,852	$1,529,412	$2,144,264	37.3	$0.0575
1996	$647,213	$1,497,793	$2,145,006	38.3	$0.0560
1997	$682,516	$1,463,041	$2,145,557	39.4	$0.0545
1998	$720,760	$1,422,337	$2,143,097	40.5	$0.0529
1999	$763,711	$1,378,368	$2,142,079	41.6	$0.0515
2000	$810,193	$1,330,764	$2,140,957	42.9	$0.0499
2001	$860,793	$1,279,194	$2,139,987	44.1	$0.0485
2002	$917,277	$1,224,364	$2,141,641	45.4	$0.0472
2003	$976,703	$1,165,685	$2,142,388	46.6	$0.0460
2004	$1,042,013	$1,101,869	$2,143,882	48.0	$0.0447
2005	$760,181	$1,034,096	$1,794,277	49.4	$0.0363
2006	$1,190,284	$960,983	$2,151,267	50.8	$0.0423
2007	$1,275,010	$877,065	$2,152,075	52.3	$0.0411
2008	$1,366,796	$787,118	$2,153,914	53.8	$0.0400
2009	$1,465,643	$690,628	$2,156,271	55.3	$0.0390
2010	$2,101,089	$586,660	$2,687,749	56.9	$0.0472
2011	$2,565,317	$443,786	$3,009,103	58.5	$0.0514
2012	$2,742,418	$269,344	$3,011,762	60.0	$0.0502
2013	$1,218,526	$82,860	$1,301,386	61.1	$0.0213
Total	$23,697,999	$21,995,089	$45,693,088		$0.98
NPV @ 6% discount					
1993	$11,302,194	$13,847,426	$25,149,620		$0.57

must share with existing development the CIP interest costs associated with net system improvements. Tables 4A and B show these adjustments and apportion interest costs between systemwide and growth-related facilities. However, new development must be credited for all its debt service payments, not just growth-related payments and not just principle or interest payments. The product of adjusting the gross SDC for these debt service credits is the *net* SDC per gallon that may be assessed on new development. This is shown

Table 9A Net Wastewater System Development Charge per Gallon

Calculation	Wastewater
Gross system development charge	$9.50
Less credit for outstanding debt service	
1985 Bond issue	$1.36
1991 Bond issue	$0.60
Less credit for future debt service	$3.22
Total credit for debt service	$5.18
Net wastewater system development charge per gallon	$4.32

Table 9B Net Water Facility System Development Charge per Gallon

Calculation	Water
Gross system development charge	$3.50
Less debt service credit for outstanding bonds	
1990 Series	$0.49
Less debt service credit for future bonds	$0.57
Total credit for debt service	$1.06
Net water facility system development charge per gallon	$2.44

in Table 9A for wastewater facilities and in Table 9B for water facilities.

SYSTEM DEVELOPMENT CHARGE SCHEDULES

We are now ready to create SDC schedules for wastewater and water facilities. A common measure of new customers must be used to determine the service demands that will result from the new connection. This characteristic may be any logical and technically defensible basis, such as water meter size, acreage, building square footage, or other measure. In choosing the common demand characteristic, a balance must be struck among the goals of: (1) using readily available data to project demand; (2) matching actual demand of a project to the SDC paid; and (3) achieving administrative ease in assessing the SDC.

Water meter size is often selected as the best available measure of average daily demand for both water and wastewater customers. The size of water meter is a generally good indicator of water demand because its physical design constrains the upper limits of demand from a particular connection. Moreover, water meters are maintained and controlled by the local government utility, allowing the monitoring of the accuracy of meter sizing. The utility can require any necessary replacement of meters that can be shown to have been sized too small for a development, and collect additional SDCs required by the change of meters.

Water usage, in turn, is related to wastewater generation. Of course, not all water that is used is returned to the wastewater system. Metcalf and Eddy, Inc. (1979), recommend assuming 80% return of water to the wastewater system for single family residential uses, and 95% for industrial and commercial (including attached housing) uses. These ratios can be used to estimate wastewater generation from water consumption. Sometimes, however, 100% return, or higher, may be used to help account for inflow and infiltration (I/I) into the wastewater system. (Indeed, if capital expenses are incurred to reduce I/I and thereby improve overall wastewater system capacity, those expenses may be recovered from system development charges.)

Since the use of such an average ratio may result in certain consumptive water users (restaurants, car washes, canneries, etc.) being undercharged for wastewater demand, the *rational nexus* approach requires that such customers be evaluated separately with their SDCs adjusted accordingly.

Another consideration is that some meters are used either exclusively or partially for fire flow. Usually, it is not appropriate to charge an SDC for fire flow capacity because such capacity is rarely if ever used, and fire protection is a communitywide benefit as well as a project-specific benefit. For this reason, fire flow capacity meters should be excluded from SDC assessments and are excluded in this example.

To establish local consumption characteristics, Columbia's 1992 water billing data were examined to determine average daily usage by water meter size. For the smallest water meter, which is used by most single-family residences, a 100% sample showed water consumption of 313 gallons per day (gpd) on an average daily basis. For larger meters, average daily water demand was estimated based on the ratio of the meter capacity to the capacity of this *equivalent residential unit* (ERU), or the $^5/_8$-in. meter. However, Columbia is working to reduce water consumption through conservation and removal of leakage in lines. It has adopted a level of service of 280 gpd/ERU, and it has based all its projections of future demand on this level of service. Wastewater demand is estimated at 80% of water usage.

At this point, the City is faced with alternative methods of SDC schedules. It has established two important facts: the price per gallon of treatment and the gallons per ERU for both wastewater and water systems. The major alternative methods of construction SDC schedules include: (1) meter size; (2) daily gallon by type of installation; (3) ERU by type of installation; and (4) fixture unit. Each approach is demonstrated in Tables 10 through 13.

Meter Size

Under this method, the customer's meter size is used as the basis for assessing both wastewater and water SDCs. The charge will vary by the ratio of the potential demand using the single residential connection, the $^5/_8$-in. meter, as the common denominator. The prospective SDC schedule is shown in Table 10.

There is a minor disadvantage with this approach. At the larger meter sizes, there can be wide ranges in actual consumption. A restaurant with 2-in. connection will consume vastly more water than an office building with the same meter connection. There are two possible solutions. The first is to allow case-by-case assessments of the nature of the actual building to adjust the meter-based SDC; but the second is more common. It assumes that most, if not all, structures connected to a meter can have interchangeable uses over time, resulting in the structures sometimes consuming relatively little but at other times consuming considerably more than the average consumption amount assumed for any given meter size. Probably for this reason, the meter size approach is the most common method of calculating SDCs calculation and assessment.

Daily Gallon and ERU by Installation Type

These alternatives are similar except that one variation is based on the average daily consumption per gallon of each unit of impact

Table 10A Water System Development Charges Based on Meter Size

Water meter (in.) size	AWWA Demand ratio[a]	Average daily water flow @ 5/8″ = 280 GPD	Water SDC @ $2.44/gal
0.625	1.0	280[b]	$683
0.750	1.1	308	$752
1.000	1.4	392	$956
1.500	1.8	504	$1,230
2.000	2.9	812	$1,981
3.000	11.0	3,080	$7,515
4.000	14.0	3,920	$9,565
6.000	21.0	5,880	$14,347
8.000[c]	29.0	8,120	$19,813

[a] American Water Works Association, Manual M-1. These ratios are also used as Equivalent Residential Units.

[b] Equivalent residential unit (ERU) is based on 3/4-in. meter size.

[c] Independent evaluations are needed for meter sizes in excess of 8 in.

associated with specific installations, and the other is based on the ERU weight for each unit of impact associated with specific installations. These approaches attempt to correct for some of the potential inequities of the meter size approach. These approaches call for each customer to be evaluated with respect to their class of customers. For example, while offices and restaurants may have the same meter connection sizes, they would be evaluated only within their particular classes of use under these approaches. Within each class, there can also be common units of impact, such as seats for restaurants and bars, square feet for offices, and bedrooms for residential.

The disadvantages of this approach are that: (1) it can be cumbersome; (2) there can still be disputes within each class; and (3) it would be difficult for local government to assess additional SDCs if a use changed from a lower to a higher level of impact in a manner that would be difficult to detect through land use permitting or business licensing.

Table 11 shows the example of the system development charges on a per gallon basis by installation type while Table 12 shows the SDCs on a per ERU basis by installation type.

Table 10B Wastewater System Development Charges Based on Meter Size

Water meter (in.) size	AWWA Demand ratio[a]	Average daily water flow @ 5/8″= 280 GPD	Average daily wastewater flow	Wastewater SDC $4.32/gal
0.625	1.0	280[b]	224[c]	$968
0.750	1.1	308	246	$1,063
1.000	1.4	392	372	$1,607
1.500	1.8	504	479	$2,069
2.000	2.9	812	771	$3,331
3.000	11.0	3,080	2,926	$12,640
4.000	14.0	3,920	3,724	$16,640
6.000	21.0	5,880	5,586	$24,132
8.000[d]	29.0	8,120	7,714	$33,324

[a] American Water Works Association, Manual M-1, p. 24. These ratios are also used as Equivalent Residential Units.

[b] Equivalent residential unit (ERU) is based on 5/8-in. meter size.

[c] Wastewater return from water consumption at 80% for 5/8- and 3/4-in. connections, which are typically detached single family dwellings, and 95% return for all other water meter connections based on Table 10A.

[d] Independent evaluations are needed for meter sizes in excess of 8 in. and for meter connections to industrial or other activities that generate exotic wastes.

Fixture Unit

The fixture unit approach is probably the most specific and thereby possibly the most equitable. Specific fixtures installed in a structure are assigned a weight, either in ERU or gallons, reflecting industry standards for expected use. Using a generally accepted plumbing code, the ERU or gallons of loading is determined for each type of fixture, such as toilet, shower, hose bib, faucet, and washing machine. The customer's SDC is assessed by multiplying the total number of fixture units by the SDC per gallon. The gallons and ERUs per fixture unit are derived from local sources using the format presented in Table 13. Table 14 shows how SDCs would be assessed using the fixture unit approach.

The disadvantage with this approach is that it requires considerable administrative sensitivity to adequately calculate the number of fixture units. Moreover, once a structure is built, it is often not practical to reassess SDCs when the customer increases fixture units, as when a homeowner may add a dishwasher without securing a permit; or even when a permit is secured, the SDC may not be assessed as a matter of administrative expediency.

Some local governments use a hybrid approach in which fixture units are the basis for calculating SDCs for residential property, while meter size is used for nonresidential development, including such quasi-residential uses as hotels, motels, resorts, and nursing homes. This approach has several advantages. First, SDCs for residential development are more sensitive to house value, and therefore, equity, since higher valued homes have more fixtures. Second, the assessments on nonresidential developments are based on average flow in meters, which assumes that while actual uses may change within a structure, on average, flow will remain the same. Third, administrative economies are realized since many local governments do not have the capacity to determine total fixtures for all developments, but do have the ability to calculate fixture units for residential development.

Table 11A Water System Development Charges Based on Per Gallon Consumption by Type of Installation

Type of installation	Unit of measure	Water flow (GPD)	Water SDC @ $2.44/gal
Airports	Per passenger	5	$12.20
	+ Employee	25	$61.00
Apartments	<3 Bedroom	180	$439.20
	3+ Bedroom	280	$683.20
	Clubhouse	500	$1,220.00
	Laundry unit	500	$1,220.00
Community halls	Per max. seat	10	$24.40
Bar, tavern	Per seat	25	$61.00
Campground	Per space	180	$439.20
Car wash	Per bay	900	$2,196.00
Church	Per seat	5	$12.20
Coin laundry	Per machine	500	$1,220.00
Commercial laundry	Per machine	800	$1,952.00
Country club	Resident member	125	$305.00
	Nonresident member	25	$61.00

Table 11A Water System Development Charges Based on Per Gallon Consumption by Type of Installation (Continued)

Type of installation	Unit of measure	Water flow (GPD)	Water SDC @ $2.44/gal
Fast food	Per seat	40	$97.60
Hospital	Per bed	250	$610.00
Light industrial	Per employee	25	$61.00
	Per shower	40	$97.60
Motel, hotel	Per unit w/restaurant	125	$305.00
	Per unit wo/restaurant	90	$219.60
Nursing home	Per bed	150	$366.00
	Per employee	25	$61.00
Office	Per 1000 sq. ft.	125	$305.00
	Per employee	25	$61.00
Police/fire station with kitchen	Per resident employee	90	$219.60
	Per nonresident employee	25	$61.00
Residence, single family	Per unit	280	$683.20
Restaurant	Per seat	5	$12.20
	Per employee	25	$61.00
Retail	Per 1000 sq.ft.	60	$146.40
School	Per student	25	$61.00
	Cafeteria seat	5	$12.20
	Gymnasium seat	5	$12.20
Service station	Per daily car	12	$29.28
	Per employee	25	$61.00
Shopping center	Per 1000 sq. ft.	125	$305.00
Swimming pool	Per swimmer	25	$61.00
	Per employee	32	$78.08
Theater	Per seat	5	$12.20
Theater, drive-in	Per space	10	$24.40
Warehouse	Per 1000 sq. ft.	60	$146.40

Note: Adapted from Public Health Service, Environmental Health Planning Guide and Environmental Health Practice in Recreational Areas, U.S. Department of Health and Human Services. No warranty is made that the figures reported in this table reflect the consumption characteristics of any given local government.

Table 11B Wastewater System Development Charges Based on Per Gallon Consumption by Type of Installation[a] (Continued)

Type of installation	Unit of measure	WW flow (GPD)[b]	WW SDC @ $4.32/gal
Airports	Per passenger	5	$21.60
	+ Employee	24	$103.68
Apartments	<3 Bedroom	162	$699.84
	3+ Bedroom	252	$1,088.64
	Clubhouse	400	$1,728.00
	Laundry unit	475	$2,052.00
Community halls	Per max. seat	10	$43.20
Bar, tavern	Per seat	24	$103.68
Campground	Per space	144	$622.08
Car wash	Per bay	760	$3,283.60
Church	Per seat	5	$21.60
Coin laundry	Per machine	475	$2,052.00
Commercial laundry	Per machine	760	$3,283.20
Country club	Resident member	100	$432.00
	Nonresident member	20	$86.40
Fast food	Per seat	38	$164.16
Hospital	Per bed	238	$1,028.16
Light industrial	Per employee	24	$103.68
	Per shower	38	$164.16
Motel, hotel	Per unit w/restaurant	119	$514.08
	Per unit wo/restaurant	86	$371.52
Nursing home	Per bed	143	$617.76
	Per employee	24	$103.68
Office	Per 1000 sq.ft.	119	$514.08
	Per employee	24	$103.68
Police/fire station	Per res. employee	86	$371.52
with kitchen	Per nonres. employee	24	$103.68
Residence, single family	Per unit	224	$967.68
Restaurant	Per seat	5	$21.60
	Per employee	24	$103.68
Retail	Per 1000 sq. ft.	572	$46.24
School	Per student	24	$103.68
	Cafeteria seat	5	$21.60

Table 11B Wastewater System Development Charges Based on Per Gallon Consumption by Type of Installation[a] (Continued)

Type of installation	Unit of measure	WW flow (GPD)[b]	WW SDC @ $4.32/gal
	Gymnasium seat	5	$21.60
Service station	Per daily car	11	$47.52
	Per employee	24	$103.68
Shopping center	Per 1000 sq.ft.	119	$514.08
Swimming pool	Per swimmer	24	$103.68
	Per employee	30	$129.60
Theater	Per seat	5	$21.60
Theater, drive-in	Per space	10	$43.20
Warehouse	Per 1000 sq. ft.	57	$226.24

[a] Adapted from Public Health Service, Environmental Health Planning Guide and Environmental Health Practice in Recreational Areas, U.S. Department of Health and Human Services. No warranty is made that the figures reported in this table reflect the consumption characteristics of any given local government.

[b] Wastewater flow is based on approximately 80% of water flow for single family residential, community and country clubs, and campground uses, 90% for apartment use, and 95% for all other uses based on Table 11A. No warranty is made that this table reflects the consumption characteristics of any given community.

Table 12A Water System Development Charges Based on ERU Consumption by Type of Installation

Type of installation	Unit of measure	EFU factor[a]	Water SDC @ $683/ERU
Airports	Per passenger	0.0179	$12.23
	+ Employee	0.0893	$60.99
Apartments	<3 Bedroom	0.6429	$439.10
	3+ Bedroom	1.0000	$683.00
	Clubhouse	1.7857	$1,219.63
	Laundry unit	1.7857	$1,219.63
Community halls	Per maximum seat	0.0357	$24.38
Bar, tavern	Per seat	0.0893	$60.99
Campground	Per space	0.6429	$439.10
Car wash	Per bay	3.2143	$2,195.37
Church	Per seat	0.0179	$12.23
Coin laundry	Per machine	1.7857	$1,219.63

**Table 12A Water System Development Charges Based on
ERU Consumption by Type of Installation (Continued)**

Type of installation	Unit of measure	EFU factor[a]	Water SDC @ $683/ERU
Commercial laundry	Per machine	2.8571	$1,951.40
Country club	Resident member	0.4464	$304.89
	Nonresident member	0.0893	$60.99
Fast food	Per seat	0.1429	$97.60
Hospital	Per bed	0.8929	$609.85
Light industrial	Per employee	0.0893	$60.99
	Per shower	0.1429	$97.60
Motel, hotel	Per unit w/restaurant	0.4464	$304.89
	Per unit wo/restaurant	0.3214	$219.52
Nursing home	Per bed	0.5357	$365.88
	Per employee	0.0893	$60.99
Office	Per 1000 sq.ft.	0.4464	$304.89
	Per employee	0.0893	$60.99
Police/fire station	Per resident employee	0.3214	$219.52
with kitchen	Per resident employee	0.0893	$60.99
Residence, single family	Per unit	1.0000	$683.00
Restaurant	Per seat	0.0179	$12.23
	Per employee	0.0893	$60.99
Retail	Per 1000 sq.ft.	0.2143	$146.37
School	Per student	0.0893	$60.99
	Cafeteria seat	0.0179	$12.23
	Gymnasium seat	0.0179	$12.23
Service station	Per daily car	0.0429	$29.30
	Per employee	0.0893	$60.99
Shopping center	Per 1000 sq.ft.	0.4464	$304.89
Swimming pool	Per swimmer	0.0893	$60.99
	Per employee	0.1143	$78.07
Theater	Per seat	0.0179	$12.23
Theater, drive-in	Per space	0.0357	$24.83
Warehouse	Per 1000 sq.ft.	0.2143	$146.37

[a] This column is based on consumption figures from Table 11A divided by water facility ERU. No warranty is made that the figures reported in this table reflect the consumption characteristics of any given local government.

Table 12B Wastewater System Development Charges Based on ERU Consumption by Type of Installation

Type of installation	Unit of measure	EFU factor[a]	WW SDC @ $968/ERU
Airports	Per passenger	0.0213	$20.58
	+ Employee	0.1060	$102.65
Apartments	<3 Bedroom	0.7233	$700.12
	3+ Bedroom	1.0000	$968.00
	clubhouse	1.7857	$1,728.56
	Laundry unit	2.1205	$2,052.66
Community halls	Per Maximum seat	0.0424	$41.04
Bar, tavern	Per seat	0.1060	$102.65
Campground	Per space	0.6429	$622.33
Car wash	Per bay	3.8170	$3,694.84
Church	Per seat	0.0213	$20.58
Coin laundry	Per machine	2.1205	$2,052.66
Commercial laundry	Per machine	3.3928	$3,284.24
Country club	Resident member	0.4464	$432.12
	Nonresident member	0.0893	$86.44
Fast food	Per seat	0.1697	$164.26
Hospital	Per bed	1.0603	$1,026.39
Light industrial	Per employee	0.1060	$102.65
	Per shower	0.1697	$164.26
Motel, hotel	Per unit w/restaurant	0.5301	$513.14
	Per unit wo/restaurant	0.3817	$369.45
Nursing home	Per bed	0.6361	$615.79
	Per employee	0.1060	$102.65
Office	Per 1000 sq.ft.	0.5301	$513.14
	Per employee	0.1060	$102.65
Police/fire station	Per res. employee	0.3817	$369.45
with kitchen	Per res. employee	0.1060	$102.65
Residence, single family	Per unit	1.0000	$968.00
Restaurant	Per seat	0.0213	$20.58
	Per employee	0.1060	$102.65
Retail	Per 1000 sq.ft.	0.2545	$246.34
School	Per student	0.1060	$102.65
	Cafeteria seat	0.0213	$20.58

Table 12B Wastewater System Development Charges Based on ERU Consumption by Type of Installation (Continued)

Type of installation	Unit of measure	EFU factor[a]	WW SDC @ $968/ERU
	Gymnasium seat	0.0213	$20.58
Service station	Per Daily car	0.0509	$49.31
	Per employee	0.1060	$102.65
Shopping center	Per 1000 sq.ft.	0.5301	$513.14
Swimming pool	Per swimmer	0.1060	$102.65
	Per employee	0.1357	$131.39
Theater	Per seat	0.0213	$20.58
Theater, drive-in	Per space	0.0424	$41.04
Warehouse	Per 1000 sq.ft.	0.2545	$246.34

[a] This column is based on ERU consumption figures from Table 12A multiplied by 0.95/0.80 for most uses except those with 80 and 90% wastewater return from water (see Table 11B). This adjustment normalizes ERU consumption figures for water from Table 12A to be proportionate to single family residential wastewater demands. No warranty is made that the figures reported in this table reflect the consumption characteristics of any given community.

Table 13A Water Fixture Units Per Equivalent Residential Connection

Fixture type	Fixture value[a]	ECUS/fixture[c]
Water related	1.00	0.0200
Bathtub	8.00	.0.1600
Bedpan washers	10.00	0.2000
Combo sink & tray	3.00	0.0600
Dental unit	1.00	0.0200
Dental lavatory	2.00	0.0400
Drinking fountain		
cooler	1.00	0.0200
public	2.00	0.0400
Kitchen sink		
$1/2''$	3.00	0.0600
$3/4''$	7.00	0.1400
Lavatory		
$3/8''$	2.00	0.0400

Table 13A Water Fixture Units Per Equivalent
Residential Connection (Continued)

Fixture type	Fixture value[a]	ERUs/ fixture[c]
$1/2''$	4.00	0.0800
Laundry tray		
$1/2''$	3.00	0.0600
$3/4''$	7.00	0.1400
Shower head	4.00	0.0800
Service sink		
$1/2''$	3.00	0.0600
$3/4''$	7.00	0.1400
Urinal, wall or stall	35.00	0.7000
Urinal, trough	12.00	0.2400
Wash sink,		
each facet set	4.00	0.0800
Water closet		
Flush valve	35.00	0.7000
Tank type	3.00	0.0600
Dishwasher		
$1/2''$	4.00	0.0800
$3/4''$	10.00	0.2000
Washing machine		
$1/2''$	5.00	0.1000
$3/4''$	12.00	0.2400
$1''$	25.00	0.5000
Other (water only)		
Wash down hoses		
$1/2''$	6.00	0.1200
$3/4''$	10.00	0.2000
Wash down hoses, 50′		
$1/2''$	6.00	0.1200
$5/8''$	9.00	0.1800
$3/4''$	12.00	0.2400
Irrigation		
(per head/100 sq.ft.)		
Spray head	1.04	0.0208
Rotary	0.26	0.0052
Snap heads	0.26	0.00
Water related[b]	46.0	
Miscellaneous factor	4.0	

Table 13A Water Fixture Units Per Equivalent Residential Connection (Continued)

Fixture type	Fixture value[a]	ERUs/ fixture[c]
Total water related[b]	50.0	
Water GPD/ERU	280.0	
Water GPD/fixture	5.60	
ERUs/Fixture	0.0200	

[a] American Water Works Association, Manual of Water Supply Practices: Sizing Service Lines and Meters, AWWA Manual No. 22, Table 4.3, p. 30. Fixtures values are based on 35 psi at the water meter outlet.

[b] Number of fixtures for typical single family home representing an equivalent residential unit connected by a $5/_8$-in. meter. Based on local information.

[c] Fixture value divided by total fixture units for typical single family home representing an equivalent residential unit connected by a $5/_8$-in. meter.

Table 13B Wastewater Fixture Units Per Equivalent Residential Connection

Fixture type	Fixture value[a]	ERUs/ fixture[c]
Water related	1.00	0.0263
Bathtub	8.00	0.2105
Bedpan washers	10.00	0.2632
Combo sink & tray	3.00	0.0789
Dental unit	1.00	0.0263
Dental lavatory	2.00	0.0526
Drinking fountain		
Cooler	1.00	0.0263
Public	2.00	0.0526
Kitchen sink		
$1/_2''$	3.00	0.0789
$3/_4''$	7.00	0.1842
Lavatory		
$3/_8''$	2.00	0.0526
$1/_2''$	4.00	0.1053

**Table 13B Wastewater Fixture Units Per Equivalent
Residential Connection (Continued)**

Fixture type	Fixture value[a]	ERUs/ fixture[c]
Laundry tray		
$1/2''$	3.00	0.0789
$3/4''$	7.00	0.1842
Shower head	4.00	0.1053
Service sink		
$1/2''$	3.00	0.0789
$3/4''$	7.00	0.1842
Urinal, wall or stall	35.00	0.9211
Urinal, trough	12.00	0.3158
Wash sink,		
each facet set	4.00	0.1053
Water closet		
Flush valve	35.00	0.9211
Tank type	3.00	0.0789
Dishwasher		
$1/2''$	4.00	0.1053
$3/4''$	10.00	0.2632
Washing machine		
$1/2$	5.00	0.1316
$3/4''$	12.00	0.3158
$1''$	25.00	0.6579
Wastewater related[b]	34.0	
Miscellaneous factor	4.0	
Total water related[b]	38.0	
Wastewater GPD/ERU	224.0	
Wastewater GPD/fixture	5.89	
ERUs/fixture unit	0.0263	

[a] American Water Works Association, Manual of Water Supply Practices: Sizing Service Lines and Meters, AWWA Manual No. 22, Table 4.3, p. 30. Fixtures values are based on 35 psi at the water meter outlet.

[b] Number of fixtures for typical single family home representing an equivalent residential unit connected by a $5/8$-in. meter. Based on local information.

[c] Fixture value divided by total fixture units for typical single family home representing an equivalent residential unit connected by a $5/8$-in. meter.

**Table 14A Water System Development Charges
Based on Fixture Units**

Fixture type	Fixture value[a]	Water use @ 5.60 GPD/fixture[b]	Water SDC @ $2.44/GPD
Water related	1.00	5.60	$13.66
Bathtub	8.00	44.80	$109.31
Bedpan washers	10.00	56.00	$136.64
Combo sink & tray	3.00	16.80	$40.99
Dental unit	1.00	5.60	$13.66
Dental lavatory	2.00	11.20	$27.33
Drinking fountain			
Cooler	1.00	5.60	$13.66
Public	2.00	11.20	$27.33
Kitchen sink			
$1/2''$	3.00	16.80	$40.99
$3/4''$	7.00	39.20	$95.65
Lavatory			
$3/8''$	2.00	11.20	$27.33
$1/2''$	4.00	22.40	$54.66
Laundry tray			
$1/2''$	3.00	16.80	$40.99
$3/4''$	7.00	39.20	$95.65
Shower head	4.00	22.40	$54.66
Service sink			
$1/2''$	3.00	16.80	$40.99
$3/4''$	7.00	39.20	$95.65
Urinal, wall or stall	35.00	196.00	$478.24
Urinal, trough	12.00	67.20	$163.97
Wash sink, each facet set	4.00	22.40	$54.66
Water closet			
Flush valve	35.00	196.00	$478.24
Tank type	3.00	16.80	$40.99
Dishwasher			
$1/2''$	4.00	22.40	$54.66
$3/4''$	10.00	56.00	$136.64
Washing machine			
$1/2''$	5.00	28.00	$68.32
$3/4''$	12.00	67.20	$163.97

Table 14A Water System Development Charges
Based on Fixture Units (Continued)

Fixture type	Fixture value[a]	Water use @ 5.60 GPD/fixture[b]	Water SDC @ $2.44/GPD
1″	25.00	140.00	$341.60
Wash down hoses			
$^1/_2$″	6.00	33.60	$81.98
$^3/_4$″	10.00	56.00	$136.64
Wash down hoses @ 50′			
$^1/_2$″	6.00	33.60	$81.98
$^5/_8$″	9.00	50.40	$122.98
$^3/_4$″	12.00	67.20	$163.97
Irrigation			
(per head/100 sq.ft.)			
Spray head	1.04	5.82	$14.20
Rotary	0.26	1.46	$3.56
Snap heads	0.26	1.46	$3.56

[a] American Water Works Association, Manual of Water Supply Practices: Sizing Service Lines and Meters, AWWA Manual No. 22, Table 4.3, p. 30. Fixtures values are based on 35 psi at the water meter outlet.

[b] For water consumption, this column is based on 60 fixture units/ERU, and for wastewater consumption it is based on 36 fixture units/ERU. This results in higher consumption per fixture unit for wastewater than for water. However, several kinds of hose connections and irrigation plumbing are exempted from wastewater SDCs. No warranty is made that the figures reported in this table reflect the consumption characteristics of any given local government.

Table 14B Wastewater System Development Charges
Based on Fixture Units

Fixture type	Fixture value[a]	Wastewater use @ 5.89 GPD/fixture[b]	Wastewater SDC @ $4.32/GPD
Water related	1.00	5.89	$25.44
Bathtub	8.00	47.12	$203.56
Bedpan washers	10.00	58.90	$254.45
Combo sink & tray	3.00	17.67	$76.33
Dental unit	1.00	5.89	$25.44
Dental lavatory	2.00	11.78	$50.89

Table 14B Wastewater System Development Charges Based on Fixture Units (Continued)

Fixture type	Fixture value[a]	Wastewater use @ 5.89 GPD/fixture[b]	Wastewater SDC @ $4.32/GPD
Drinking fountain			
Cooler	1.00	5.89	$25.44
Public	2.00	11.78	$50.89
Kitchen sink			
$1/2''$	3.00	17.67	$76.33
$3/4''$	7.00	41.23	$178.11
Lavatory			
$3/8''$	2.00	11.78	$50.89
$1/2''$	4.00	23.56	$101.78
Laundry tray			
$1/2''$	3.00	17.67	$76.33
$3/4''$	7.00	41.23	$178.11
Shower head	4.00	23.56	$101.78
Service sink			
$1/2''$	3.00	17.67	$76.33
$3/4''$	7.00	41.23	$178.11
Urinal, wall or stall	35.00	206.15	$890.57
Urinal, trough	12.00	70.68	$305.34
Wash sink, each facet set	4.00	23.56	$101.78
Water closet			
Flush valve	35.00	206.15	$890.57
Tank type	3.00	17.67	$76.33
Dishwasher			
$1/2''$	4.00	23.56	$101.78
$3/4''$	10.00	58.90	$254.45
Washing machine			
$1/2''$	5.00	29.45	$127.22
$3/4''$	12.00	70.68	$305.34
$1''$	25.00	147.25	$636.12

Table 14B Wastewater System Development Charges
Based on Fixture Units (Continued)

Fixture type	Fixture value[a]	Wastewater use @ 5.89 GPD/fixture[b]	Wastewater SDC @ $4.32/GPD

[a] American Water Works Association, Manual of Water Supply Practices: Sizing Service Lines and Meters, AWWA Manual No. 22, Table 4.3, p. 30. Fixtures values are based on 35 psi at the water meter outlet.

[b] For wastewater consumption, this column assumes 36 fixture units/ERU. This results in higher consumption per fixture unit for wastewater than for water. However, several kinds of hose connections and irrigation plumbing are exempted from wastewater SDCs. No warranty is made that the figures reported in this table reflect the consumption characteristics of any given local government.

8   Calculating System Development Charges for Stormwater Facilities

INTRODUCTION

Stormwater runoff usually occurs during and immediately after rainfall, but can sometimes involve groundwater flow and snow melt. The primary purpose of stormwater systems and associated stormwater management policies is to control runoff in ways that minimize hazards to life and property, and minimize inconvenience to the general public. Urban development increases the frequency and severity of flooding by removing vegetation, filling natural water storage areas, covering floodplains and watersheds with pavement, and reducing the size of the natural channel available for flood flows. These disturbances increase flood velocities and flood heights. As watersheds are developed, buildings built in the floodplains face increased risk as flood levels rise. Development adjacent to natural floodplains may become subject to flooding as floodplain areas increase.

A primary consideration of stormwater analysis is amount and change in impervious surfaces. An impervious surface is any material that prevents absorption of stormwater into the ground. Retention and detention basins and dry wells allowing stormwater to percolate directly into the ground are not usually considered impervious surfaces. Graveled areas usually are. One method by which impervious surfaces can be defined is in terms of a percolation rate in minutes per inch.

Unlike wastewater and water systems, however, stormwater systems can be affected by development patterns occurring outside the jurisdiction of local government. For this reason, stormwater facility planning is best done on a drainage-basin level involving all relevant governmental units. This is not always possible, however.

Stormwater facility planning must consider hydrologic factors specific to the drainage basin and the development patterns of that basin, such as discharge rates, runoff volume, flow velocities, discharge characteristics at various stages of a rainfall or other runoff event, the condition of downstream flow conditions, and water quality. Stormwater facility planning also includes identification of system alternatives considering: land use patterns; storage, channeling and treatment options; risks to life, property, and convenience; and costs of installation, operation and maintenance, and replacement.

Stormwater facility planning is based on the magnitude of peak storm events. The magnitude of a storm is described by its likelihood of occurrence. For example, a "25-year storm" is likely to occur, on the average, once every 25 years. Another way of referring to a 25-year storm would be to say it is a storm that has a 4% chance of occurring in any single given year.

Stormwater system development charges (SDCs) are used to help finance the particular stormwater system resulting from the planning process. SDCs for stormwater facilities are conceptually similar to those of wastewater and water SDC, but have many practical differences. Similar developments in different locations can have different levels of runoff quantity and quality, depending on soils, slopes, amount of impervious surface, setting with respect to other land uses, and other factors. In addition, stormwater systems can be affected by development patterns occurring outside the jurisdiction of local government. For this reason, stormwater facility planning is best done on a drainage-basin level involving all relevant governmental units, although this is not always possible.

One of the basic questions involved in calculating stormwater SDC is the unit of development impact on which to measure the impact of any particular development. The primary purpose of this chapter is to review some of the more general ways in which common units of impact measure may be developed and applied to stormwater SDCs. The chapter also presents two general approaches to the calculation of such fees. The first approach reviewed is where a stormwater collection system is installed to serve a region composed of one major drainage basin. The second approach is the calculation of fees in lieu of construction of stormwater storage facilities to accommodate specific developments. Both fees can be combined. The calculation of stormwater SDCs follows the same steps as for wastewater and water SDCs.

REGIONAL FACILITIES APPROACH

The first approach is where a stormwater collection system is installed to serve a region composed of one major drainage basin. This approach will apply SDCs to one common unit of impact, the equivalent drainage unit or EDU.

Service Area

Stormwater becomes runoff and flows downhill in depressions and ravines. Water collects from the network of small ravines to form a stream or creek. Land area that is drained by the stream or river and its tributaries is called a "drainage basin" or "watershed". A divide or ridge line separates drainage basins. The City of Columbia is situated in a single drainage basin that drains into the Cherokee River, which empties into the Gulf of Mexico. Thus, one service area is considered.

Level of Service and Projections of Demand

The most important factor in determining the impact of a development on stormwater facility needs is the addition of impervious cover. To calculate the existing and growth-related costs per acre of impervious cover, it is necessary to estimate the amount of existing and future impervious cover for each of the two proposed service areas.

The drainage basin studies that the City has completed were designed to accommodate the city's stormwater needs to the year 2010, based on its mid-1980s planning documents. Existing (1993) and future (2010) acres of impervious cover are estimated based on the City's existing land use inventory, population and employment estimates and projections by census tract, and typical impervious cover ratios for four general land use types. The first step in the analysis is to derive average population and employment densities for the four land use types and apply those to future population and employment estimates. This has been done based on the City's current land use inventory and 1990 population and employment data. Columbia has a residential density of about 8 persons per acre, and employment densities ranging from about 8 employees per acre for industrial land to about 69 employees per acre for office/commercial development, as shown in Table 1.

The next step is to estimate existing and future developed acres by each of the four land use categories for both service areas. This has been accomplished using the population and employment densities calculated in Table 1, and population and employment estimates and projections. Population and employment for each service area is based on their approximate correlation with census tract boundaries. The results are presented in Table 1.

The final step is to estimate the amount of existing and future impervious surface in each service area. This is done multiplying the estimates of developed acres by land use category derived in the above table by impervious cover ratios, as shown in Table 2. The impervious cover ratios are based on reasonable assumptions and data from other communities.

Inventory of Existing Systems and Planned Expansions

Columbia has been installing a regional stormwater collection system for nearly 3 decades and plans to install additional facilities over the next decade. The current value of the existing stormwater system and the planned expansions apportioned for new development is determined in the same manner as determined for water and wastewater systems in Chapter 6. For brevity, Table 3 summarizes what amounts to more than 300 stormwater items by subbasin, and Table 4 summarizes nearly an equal number of additions to the system planned to be installed over the next decade. The total cost attribution method is used.

Table 5 summarizes values and costs presented in Tables 3 and 4 attributed to new development. The analysis is similar to the steps presented in Chapter 7. It shows that new development can be attributed with $0.32 per square foot of impervious surface. Beware, however, that for brevity of exposition, Table 5 does not consider the effects of revenue credits on the net fee to be assessed; the table only reports the cost per impervious square foot.

Information displayed in Table 5 must be converted into common units of assessment for SDC purposes. The most common unit of measure is the *equivalent drainage unit* (EDU), or its alternates: equivalent service unit (ESU) and equivalent residential units (ERU).

Table 1 Average Population and Employment Densities: 1990

Land use category	Population, employment 1990	Developed acres 1990	Population, employment per acre	Population, employment 2010	Developed acres 2010
Residential	319,583	39,406	8.11	355,219	43,800
Retail	40,961	1,411	29.03	49,992	1,722
Office/commercial	211,561	3,071	68.89	295,505	4,290
Industrial	58,178	7,001	8.31	74,039	8,910
Total		50,889			58,722

Table 2 Land Use by Service Area, 1993–2010

| Service area/ | Developed acres | | Imperv. cover | Imperv. acres | |
land use	1993	2010	ratio	1993	2010
Residential	39,406	43,800	0.21	8,275	9,198
Retail	1,411	1,722	0.67	945	1,154
Office/commercial	3,071	4,290	0.67	2,058	2,874
Industrial	7,001	8,910	0.50	3,501	4,455
Cherokee Basin	50,889	58,722		14,779	17,681

Another common unit is acre of development, an example of which is presented below. Table 6 calculates the stormwater SDC per EDU.

Table 7 calculates how the costs per EDU are applied to generate the cost per residential unit for single- and multiple-family residential developments, and per 1000 sq. ft. of gross leasable area for office (including government office), commercial (including institutional), and industrial developments.

Inasmuch as local governments may wish to differentiate SDCs by more refined categories than those presented in the tables above, Table 8 uses information from Table 6 to determine stormwater SDCs by zoning district classification.

Debt Service Costs and Credits

Debt service costs and credits are handled in one of three ways. The most common way is for stormwater facilities to be financed by revenue bonds retired by water or wastewater rates. Since volume of water or wastewater is used to calculate rates for a given customer, the debt service costs and credit would be handled the same as for water and wastewater revenue bonds presented in Chapter 6. Sometimes, stormwater facilities are financed from general obligation bonds. Appendix 2 shows how debt service credits are calculated when retired through property taxes. At other times, stormwater facility revenue bonds may be retired by stormwater utility rates. However, those rates are often not calibrated to volume of stormwater generated, but rather in terms of general land use categories. In these situations, one would need to develop a per-cubic-foot revenue credit by general land use similar to the credit per gallon of water and wastewater consumption shown in Chapter 6. This approach would require estimating annual

Table 3 Current Value of Existing Improvements by Subbasin

Improvement types	North Creek	South Creek	West Creek	Total
RCP pipe	$19,083,775	$172,500	$2,297,500	$21,553,775
Catch basins	$1,315,200	$6,600	$316,800	$1,638,600
Manholes	$1,800	$14,400	$230,400	$246,600
Yard inlets	$33,000	$13,500	$19,500	$66,000
Headwalls	$513,000	$44,800	$378,650	$936,450
Gabion slope protection	$1,503,072	$495,620	$377,978	$2,376,670
Rip Rap outlet protection	$28,200	$0	$0	$28,200
Concrete channel lining	$0	$0	$647,459	$647,459
Retaining wall	$0	$36,000	$1,740,250	$1,776,250
Curb and gutter	$725,813	$27,313	$303,375	$1,056,501
Raise existing curb/gutter	$0	$0	$0	$0
Private driveway bridges	$0	$0	$0	$0
Priv. driveway adjustments	$61,380	$1,980	$92,400	$155,760
Easements	$4,289,125	$470,770	$1,943,340	$6,703,235
Clearing and grubbing	$1,372,520	$172,849	$707,290	$2,252,659
Unclassified excavation	$0	$19,384	$185,196	$204,580
Repair structures	$156,000	$0	$0	$156,000
Engineering/admin.	$2,909,091	$147,572	$924,013	$3,980,676
Total current value	$31,991,976	$1,623,288	$10,164,151	$43,779,415
Growth-related share	8.75%	11.60%	0.00%	
Growth-related current value	$2,799,298	$188,301	$0	$2,987,599
Systemwide current value	$29, 192,678	$1,434,987	$10,164,151	$40,791,816

Table 4 Planned Expansions by Subbasin

Improvement types	North Creek	South Creek	West Creek	Total
RCP pipe	$456,250	$8,080,150	$16,128,135	$24,664,535
Catch basins	$77,600	$1,065,975	$648,870	$1,792,445
Manholes	$30,600	$333,200		$363,800
Yard inlets	$9,000		$26,962	$35,962
Headwalls	$56,500	$274,295	$345,067	$675,862
Gabion slope protection	$16,400	$2,313,492	$144,635	$2,474,527
Rip Rap outlet protection	$0	$0	$5,580	$5,580
Concrete channel lining	$80,377	$22,595	$0	$102,972
Retaining wall	$1,360,750	$93,995	$0	$1,454,745
Curb and gutter	$36,750	$94,750	$185,670	$317,170
Raise existing curb/gutter	$0	$0	$6,713	$6,713
Private driveway bridges	$269,842	$0	$3,808,761	$4,078,603
Private driveway adjustments	$1,320	$27,720	$19,544	$48,584
Easements	$932,225	$1,617,175	$3,968,558	$6,517,958
Clearing and grubbing	$331,458	$166,968	$1,269,338	$1,767,764
Unclassified excavation	$45,470	$198,090	$55,370	$298,930
Repair structures	$0	$0	$146,755	$146,755
Engineering/admin.	$370,454	$1,431,041	$2,675,996	$4,477,491
Total costs	$4,074,996	$15,719,446	$29,435,954	$49,230,396
Growth-related percent	0.00%	46.75%	68.40%	
Growth-related expansion costs	$0	$7,348,841	$20,134,193	$27,483,034
Systemwide expansion costs	$4,074,996	$8,370,605	$9,301,761	$21,747,362

Table 5 Stormwater Facility Values and Costs
Attributable to Growth

Assessment consideration	Calculation
Current demand, impervious acres	14,779
Projected demand, impervious acres	17,681
Projected growth-related demand, impervious acres	2,902
Current value systemwide improvements (not growth related)	$40,791,816
CIP systemwide improvements (not growth related)	$21,747,362
Total value, systemwide improvements	$62,539,178
Systemwide improvement costs, per impervious acre	$3,537.08
Systemwide improvement costs, per impervious square foot	$0.08
Current value growth-related costs	$2,987,599
CIP growth-related costs	$27,483,034
Total growth-related costs	$30,470,633
Growth-related costs, per new impervious acre	$10,499.87
Growth-related costs, per impervious square foot	$0.24
Total cost, per impervious square foot	$0.32

Table 6 Stormwater System Development
Charge per Equivalent Drainage Unit

Cost component	Service area
Building footprint (sq. ft.)	2400
Driveway (sq. ft.)	400
Patio (sq. ft.)	300
Total impervious area (sq. ft.)	3100
SDC cost per square foot	$0.32
SDC per EDU	$992

changes by general land use category, similar to the manner shown in Appendix 2 for the property tax credit.

FEE IN-LIEU OF CONSTRUCTION APPROACH

The second approach, fees in lieu of constructing on-site detention/ retention facilities, is more common and more simple, but potentially less defensible than the approach presented earlier. It is used where

Table 7 Equivalent Drainage Units per Unit of Development

Land use	Unit of impact	EDUs per unit	SDC per unit
Single family	Res. Unit	1.00	$992
Multiple family	Res. Unit	0.79	$784
Office/government	1000 sq. ft.	1.25	$1240
Commercial/institutional	1000 sq. ft.	1.30	$1290
Industrial	1000 sq. ft.	1.23	$1220

Table 8 Calculation of SDCs by Zoning District

Zoning district	Unit of impact	EDUs per unit	SDCs @ $1054/EDU
R-1	Res. Unit	1.00	$992
R-2	Res. Unit	1.00	$992
R-4	Res. Unit	1.00	$992
R-6	Res. Unit	1.00	$992
R-8	Res. Unit	1.00	$992
R-10	Res. Unit	0.79	$784
R-20	Res. Unit	0.79	$784
R-30	Res. Unit	0.79	$784
OI-1	1000 sq. ft.	1.25	$1240
OI-2	1000 sq. ft.	1.25	$1240
OI-3	1000 sq. ft.	1.25	$1240
C-1	1000 sq. ft.	1.30	$1290
C-2	1000 sq. ft.	1.30	$1290
C-3	1000 sq. ft.	1.30	$1290
I-1	1000 sq. ft.	1.23	$1220
I-2	1000 sq. ft.	1.23	$1220

individual developments are required to construct stormwater storage facilities on-site as project improvements based on the locally adopted level of service. Sometimes, these individual storage facilities may hold water and then release it into nearby streams. In other situations, the water is held until it drains into the soils. In still other situations, water is held and then released into regional transmission lines that transmit it to centralized treatment facilities. The third

Table 9 In-lieu Fees per Acre of Development by Zoning District

Zoning district	Unit of impact	Cubic feet per unit of development	In-lieu fee per cf @ $23,400 per acre-ft.	In-lieu fee per unit of development
R-1	Res. Unit	726	$0.54	$392
R-2	Res. Unit	608	$0.54	$328
R-4	Res. Unit	502	$0.54	$271
R-6	Res. Unit	502	$0.54	$271
R-8	Res. Unit	454	$0.54	$245
R-10	Res. Unit	396	$0.54	$214
R-20	Res. Unit	396	$0.54	$214
R-30	Res. Unit	396	$0.54	$214
OI-1	1000 sq. ft.	628	$0.54	$339
OI-2	1000 sq. ft.	628	$0.54	$339
OI-3	1000 sq. ft.	628	$0.54	$339
C-1	1000 sq. ft.	652	$0.54	$352
C-2	1000 sq. ft.	652	$0.54	$352
C-3	1000 sq. ft.	652	$0.54	$352
I-1	1000 sq. ft.	619	$0.54	$334
I-2	1000 sq. ft.	619	$0.54	$334

solution is becoming more common in urbanized areas. The in-lieu approach does not necessarily assume that any of the three options are locally employed. Regardless of the option, each option requires construction of an on-site storage facility; or in the absence of such a facility, a fee in-lieu to be used by local government to construct such a facility elsewhere.

We shall assume that the service area and level of service are the same as in the approach presented above. A local hydrological study estimates the cost of constructing stormwater storage facilities to average $23,400 per acre-foot. This cost can be slightly higher per acre-foot for very small facilities and can be lower per acre-foot for very large facilities. Nonetheless, the average cost may be used to determine reasonable fees in-lieu of a developer constructing such facilities. The cost per EDU can be derived from the rational method of estimating stormwater runoff.

Rather than using EDUs, however, consider the nature of fees if one used acres of land developed instead. For any class of land use, especially residential, there may be very different runoff coefficients. For example, single-family estate areas may average one home per acre, while a moderate-density, single-family area may have up to eight homes per acre. While the estate may have more impervious surface than the average home, the concentration of eight homes on 1 acre results in vastly greater impervious surface. It is thus reasonable to establish average coefficients of runoffs for each class of land use. This can be easily done using the local zoning districts as a guide, which was done by city of Columbia and shown in Table 9.

Under this method, Table 9 becomes the fee in-lieu schedule applied to development within each zoning district for situations where developers choose not to install stormwater storage facilities on-site. Consistent with *rational nexus* requirements, individual developers who demonstrate lower runoff coefficients would be assessed lower fees associated with reduced generation of acre-feet.

Appendix 1

**Engineering News and Record Construction Costs Index
and Current Value Factors 1940–1992**

Year	ENR index 1st Qtr	ENR factor 1992	ENR index 2nd Qtr	ENR factor 1992	ENR index 3rd Qtr	ENR factor 1992	ENR index 4th Qtr	ENR factor 1992
1940	238	20.7017	242	20.5496	244	20.6639	249	20.3173
1941	251	19.6295	257	19.3502	263	19.1711	266	19.0188
1942	262	18.8053	274	18.1496	282	17.8794	284	17.8134
1943	285	17.2877	290	17.1483	294	17.1497	295	17.1492
1944	295	16.7017	299	16.6321	300	16.8067	302	16.7517
1945	305	16.1541	309	16.0939	309	16.3172	313	16.1629
1946	324	15.2068	348	14.2902	360	14.0056	368	13.7473
1947	392	12.5689	403	12.3400	426	11.8357	432	11.7106
1948	438	11.2489	456	10.9057	478	10.5481	478	10.5837
1949	475	10.3726	478	10.4038	480	10.5042	480	10.5396
1950	488	10.0963	507	9.8087	530	9.5132	531	9.5273
1951	539	9.1410	543	9.1584	543	9.2855	548	9.2318
1952	551	8.9419	562	8.8488	585	8.6188	588	8.6037
1953	587	8.3935	594	8.3721	610	8.2656	611	8.2799
1954	614	8.0244	622	7.9952	640	7.8781	643	7.8678
1955	646	7.6269	656	7.5808	673	7.4918	673	7.5171
1956	680	7.2456	692	7.1864	705	7.1518	704	7.1861
1957	709	6.9492	721	6.8974	738	6.8320	738	6.8550
1958	744	6.6223	758	6.5607	773	6.5226	774	6.5362
1959	781	6.3086	795	6.2553	813	6.2017	809	6.2534
1960	813	6.0603	827	6.0133	830	6.0747	831	6.0878
1961	834	5.9077	852	5.8369	853	5.9109	855	5.9170
1962	863	5.7092	877	5.6705	880	5.7295	881	5.7423
1963	884	5.5735	899	5.5317	914	5.5164	915	5.5290
1964	922	5.3438	935	5.3187	947	5.3242	948	5.3365
1965	958	5.1430	969	5.1321	986	5.1136	988	5.1204
1966	998	4.9369	1029	4.8328	1034	4.8762	1034	4.8926
1967	1043	4.7239	1068	4.6564	1092	4.6172	1098	4.6075
1968	1117	4.4109	1154	4.3094	1186	4.2513	1201	4.2123

**Engineering News and Record Construction Costs Index
and Current Value Factors 1940–1992 (Continued)**

Year	ENR index 1st Qtr	ENR factor 1992	ENR index 2nd Qtr	ENR factor 1992	ENR index 3rd Qtr	ENR factor 1992	ENR index 4th Qtr	ENR factor 1992
1969	1238	3.9798	1270	3.9157	1285	3.9237	1305	3.8766
1970	1314	3.7496	1375	3.6167	1421	3.5482	1445	3.5010
1971	1496	3.2934	1589	3.1296	1654	3.0484	1672	3.0257
1972	1697	2.9034	1761	2.8240	1786	2.8231	1816	2.7858
1973	1859	2.6503	196	25.3724	1929	2.6138	1939	2.6091
1974	1940	2.5397	1993	2.4952	2089	2.4136	2101	2.4079
1975	2128	2.3153	2205	2.2553	2275	2.2163	2297	2.2024
1976	2322	2.1219	2410	2.0635	2465	2.0454	2490	2.0317
1977	2513	1.9606	2541	1.9571	2644	1.9070	2660	1.9019
1978	2693	1.8296	2753	1.8064	2851	1.7685	2869	1.7633
1979	2886	1.7072	2984	1.6666	3120	1.6160	3140	1.611
1980	3159	1.5597	3198	1.5550	3319	1.5191	3376	1.4985
1981	3384	1.4560	3496	1.4225	3657	1.3787	3695	1.3691
1982	3721	1.3241	3815	1.3035	3902	1.2922	3950	1.2808
1983	4006	1.2299	4073	1.2210	4142	1.2173	4110	1.2309
1984	4118	1.1965	4161	1.1951	4176	1.2074	4144	1.2208
1985	4151	1.1869	4201	1.1838	4229	1.1922	4228	1.1965
1986	4231	1.1645	4303	1.1557	4335	1.1631	4351	1.1627
1987	4359	1.1303	4387	1.1336	4456	1.1315	4478	1.1297
1988	4484	1.0988	4525	1.0990	4535	1.1118	4568	1.1075
1989	4574	1.0772	4599	1.0813	4658	1.0824	4685	1.0798
1990	4691	1.0503	4732	1.0509	4774	1.0561	4777	1.0590
1991	4772	1.0325	4818	1.0322	4891	1.0309	4889	1.0348
1992	4927	1.0000	4973	1.0000	5042	1.0000	5059	1.0000

Note: The ENR Construction Cost Index is based on a 20 U.S. city average cost of a hypothetical unit of construction requiring: 6 barrels of portland cement; 1088 thousand board feet of 2″ × 4″ lumber; 2500 pounds of structural steel; 68.38 h of skilled labor; and 200 h of common labor. The ENR Construction Cost Index is published monthly. This table should be updated annually.

Appendix 2

Calculation of System Development Charge Credits Attributable to Past and Future General Obligation Bond Debt Service

INTRODUCTION

Oftentimes general obligation (G.O.) bonds are issued to pay for water, wastewater, or stormwater facilities. Those bonds are usually secured by the real property of the sponsoring local government and retired through property tax assessments against all taxable real property. New development will pay property taxes that will be used in part to retire the debt incurred to pay for the facilities it may also pay for through system development charges (SDCs). Thus, it should be entitled to a credit equal to the average present value of the stream of property tax payments it may make over the remaining life of the bond after its developed value is placed on the tax digest. However, since vacant land also pays property taxes used to retire this debt, a credit for past payments is also due.

There is a simple way to make these credit adjustments. If a local government sells G.O. bonds to pay for, say, 50% of the cost of new facilities, it may substract this amount from SDCs. In this manner, full credit is given because only the portion of the facilities financed from other than G.O. bonds forms the basis for calculating SDCs. It would be true that new development would nevertheless make property tax payments to retire those G.O. bonds, but its SDCs would not have included the facilities financed from the bonds. But this also means that existing development will subsidize new development because it is *all* real property, not just new development, that retires G.O. bonds sold to expand facilities for new development.

The purpose of this appendix is to demonstrate how credits can be reasonably calculated for future and past property tax payments used to retire G.O. bonds for the same facilities for which new development is assessed SDCs. Seven steps are involved:

1. Determining current assessed value of real property.
2. Projecting annual average increases in the assessed value base of the local government over the life of the G.O. bond.
3. Determining annual debt service payments per $1000 assessed value and determining the net present value of the stream of those payments from the time new development's value is assessed and the year in which the bond is retired.
4. Determining the credit attributable to general land use types based on average unit values of those types.
5. Determining total payments per $1000 valuation made by vacant land in the past.
6. Determining the credits for past debt service payments by general land use type.
7. Summing credits for future and past debt service payments by general land use type.

These steps will be illustrated through an example involving a local government issuance of $29,440,000 in general obligation bonds for stormwater facilities, issued in 1988 but with payments commencing in 1989 and ending in 2008.

Determining Current Assessed Value of Real Property

The first step is collecting information on the total assessed valuation of the local government for every year in which debt service has been paid in the past. This information is shown for Columbia in Table 1.

Projecting Annual Average Increases in Assessed Value

This step involves a simple extrapolation of current assessed valuation per resident and worker in the city to a future date. Since population and employment projections of the local comprehensive

**Table 1 Current Assessed Value
of Taxable Real Property**

Year	Assessed value
1989	$11,068,416
1990	$12,223,896
1991	$12,541,704
1992	$12,957,774

plan use 1990 and 2010 as the projection horizon, Table 2 uses those years as well. The extrapolation involves determining the average assessed value for residents and workers combined for the base year, 1990, and then multiplying that figure by the combined resident and worker projections for 2010. More complex projections were not engaged, principally because local government property tax data in this situation were not refined enough to allow for it. One could project assessed valuation for general land use types if one had data on current and projected residents and workers by land use type, and current assessed values by the same land use types.

Determining Future Debt Service Credit per $1000 Assessed Value

Table 3 shows how the average annual change in assessed value is used to project future debt service payments per $1000 valuation (in current dollars), and then how the stream of those payments are discounted to the present. Notice that for years in which actual assessed values are known, 1989–1992 in this example, those values are used instead of projected values. Otherwise, assessed value estimations of future years are based on straight extrapolations. The average annual change is assessed value should probably be recalculated every 5 years.

Determining the Future Debt Service Credit by General Land Use Type

The future debt service credit per $1000 assessed valuation derived from Table 3 is applied to general land use types as shown in Table 4. The credit is applied to the average value of new construction based on data provided by the local government office of assessment

Table 2 Projecting Annual Average Assessed Value Growth

Base	Population & employment 1990	Assessed value 1990	Assessed value per capita	Population & employment 2010	Assessed value 2010
Population	168,210			237,685	
Employment	53,808			76,032	
Population + employment	222,018	$12,223,896	$55,058	313,717	$17,272,631
Average annual change in assessed value ($)					$252,437

Table 3 Debt Service on General Obligation Bond and Present Value of Future
Debt Service Payments per $1000 Valuation

Year	Principal payment	Interest payment	Total payment	Assessed value	Payment per $1000 value
1988	$0	$0	$0		
1989	$190,000	$2,488,567	$2,678,567	$11,068,416	$0.2420
1990	$375,000	$2,117,583	$2,492,583	$12,223,896	$0.2039
1991	$535,000	$2,092,665	$2,627,665	$12,541,704	$0.2095
1992	$650,000	$2,058,673	$2,708,673	$12,957,774	$0.2090
1993	$825,000	$2,014,498	$2,839,498	$13,210,211	$0.2149
1994	$925,000	$1,960,198	$2,885,198	$13,462,648	$0.2143
1995	$1,015,000	$1,898,073	$2,913,073	$13,715,085	$0.2124
1996	$1,080,000	$1,828,635	$2,908,635	$13,967,522	$0.2082
1997	$1,160,000	$1,752,165	$2,912,165	$14,219,959	$0.2048
1998	$1,240,000	$1,668,745	$2,908,745	$14,472,396	$0.2010
1999	$1,415,000	$1,575,113	$2,990,113	$14,724,833	$0.2031
2000	$1,790,000	$1,460,440	$3,250,440	$14,977,270	$0.2170
2001	$1,925,000	$1,325,738	$3,250,738	$15,229,707	$0.2134
2002	$2,075,000	$1,178,700	$3,253,700	$15,482,144	$0.2102
2003	$2,235,000	$1,018,113	$3,253,113	$15,734,581	$0.2067
2004	$2,410,000	$842,720	$3,252,720	$15,987,018	$0.2035

Table 3 Debt Service on General Obligation Bond and Present Value of Future Debt Service Payments per $1000 Valuation (Continued)

Year	Principal payment	Interest payment	Total payment	Assessed value	Payment per $1000 value
2005	$2,600,000	$651,040	$3,251,040	$16,239,455	$0.2002
2006	$2,810,000	$441,350	$3,251,350	$16,491,892	$0.1971
2007	$3,040,000	$211,680	$3,251,680	$16,744,329	$0.1942
2008	$1,145,000	$45,800	$1,190,800	$16,996,766	$0.0701
Totals	$29,440,000	$28,630,496	$58,070,496		$4.0355
NPV @ 6% discount, 1993 dollars	$15,968,516	$14,141,521	$30,110,037		$2.05

Table 4 Credit for Future G.O. Bond Debt Service

Land use	Unit of impact	Average value, new development 1992	Credit/ $1000 value	Credit/ unit of develop.
Single family	Res. unit	$135,000	$2.05	$276.75
Multiple family	Res. unit	$55,000	$2.05	$112.75
Hotel/motel	Room	$67,500	$2.05	$138.38
Office	1000 sq. ft.	$162,500	$2.05	$333.12
Retail	1000 sq. ft.	$142,000	$2.05	$291.10
Industry	1000 sq. ft.	$57,500	$2.05	$117.87

and taxation. Individual developers may wish to claim higher credits if their values are higher than the average value used for credit calculations.

However, except for possibly highly unusual cases, allowing higher credits should be discouraged for several reasons. First, if higher credits are allowed for some developments, and since an average value is used for all developments including lower-valued ones, the total credits allowed by local government would be higher than expected. Existing taxpayers would thereby be contributing more to debt service as a consequence.

Second, and perhaps more important, the SDC itself is usually based on average consumption figures based on meter size, ERU, or general land use classifications. A higher-valued development, such as estate homes, is paying a SDC based only on the average consumption patterns of the community, yet larger homes will consume more water and wastewater and usually generate greater runoff than the average home. Thus, if a higher credit is sought by a developer, the local officials may need to calculate the additional demands placed on facilities attributable to such development withthe SDCs adjusted accordingly.

Determining Credit For Past Debt Service Payments

Table 5 illustrates the calculation of credits for past debt service payments. Since the debt service payments per $1000 valuation are determined in Table 4, those values are reported for each year of past debt service and then summed to arrive at a total credit per $1000 valuation of vacant land.

**Table 5 Credit per $1000
Valuation for Past Debt Service**

Year	Past debt service per $1000 value
1989	$0.2420
1990	$0.2039
1991	$0.2095
1992	$0.2090
Total	$0.8644

Table 6 Credit for Past Debt Service by General Land Use

Land use	Unit of impact	Average value, new development 1992	Average developed land to value ratio	Credit @ $0.8644/ $1000 value
Single family	Res.unit	$135,000	25.00%	$29.17
Multiple family	Res. unit	$55,000	22.50%	$10.70
Hotel/motel	Room	$67,500	20.00%	$11.67
Office	1000 sq. ft.	$162,500	17.50%	$24.58
Retail	1000 sq. ft.	$142,000	17.50%	$21.48
Industry	1000 sq. ft.	$57,500	15.00%	$7.46

Determining Past Debt Service Credits by General Land Use Type

The information from Table 5 is applied to the same general land use types used in Table 4 for credit calculation purposes. Since land value is only a portion of the total value of a unit of development, the land value share must be broken out. In this example, the percent of total developed value per unit attributable to land is determined from the local property tax assessment office and is based on a sample of developed properties. Most property assessment offices divide assessed value into "land" and "improvements". The credit information from Table 5 is applied to the land value portion of developed property, which is shown in Table 6.

Table 7 Credit Summary for G.O. Bond Debt Service in 1993

Land use	Unit of impact	Land value credit for past payments	Developed unit credit for future payments	Total tax credits
Single family	Res. unit	$29.17	$276.75	$305.92
Multiple family	Res. unit	$10.70	$112.75	$123.45
Hotel/motel	Room	$11.67	$138.38	$150.05
Office	1000 sq. ft.	$24.58	$333.12	$357.70
Retail	1000 sq. ft.	$21.48	$291.10	$312.58
Industry	1000 sq. ft.	$7.46	$117.87	$125.33

Two interesting observations are made. First, the time-value of money is accounted for because the credit is applied to the average value of a unit of land in the current year. Second, the land value credit is based on percent of developed property value attributable to land. This will tend to overstate credits for past payments to the extent that raw, undeveloped land is valued on the tax digest at less than the amounts implied in this credit calculation process. Local officials may wish to further adjust the average land-to-developed-value ratio to account for this.

Total Debt Service Credits by General Land Use Type

Table 7 shows the total debt service credits attributable to new development by general land use type. The credit calculations should be adjusted annually to keep values current. The practical effect of not adjusting values is probably to award higher credits to new development since, for each successive year, the credit for future debt service payments will fall by more than the credit for past debt service payments.

Appendix 3
Adjustments to Meter Size Ratios to Account for Peak Demand

Although calculating and assessing water and wastewater system development charges (SDCs) is most commonly based on average daily use, usually measured for a typical month or averaged over a year, sometimes they may be based on peak demand. If so, two important considerations must be addressed. First, there are three kinds of meters through which water flows and each measures volume differently. Second, if the peak demand factor must be used, the system capacity must be adjusted from average daily rating to peak demand rating. This second consideration is highly variable, depending on the nature of storage facilities, treatment production capacity, the extent to which stormwater flows into wastewater systems and is considered in rated capacity, and so forth. The second consideration must be locally determined. It is the first consideration that this appendix addresses.

Water meters can differ not only in size, but in type. The standard residential meter uses a disk that is rotated by the passage of water to measure its flow and is the most reliable way to measure flow at low volumes. In contrast, a turbine meter uses a turbine instead of a disk and is the most reliable way to measure continuous high-volume flows. Customers whose use over time varies between normal and high-flow volumes install compound meters that switch between disk and turbine meters as the flow dictates. The meter capacity for each type of meter is shown in Table 1.

Table 1 Meter Capacities by Meter Size and Type

Meter size and type	Meter capacity[a]
$5/8 \times 3/4$-in. Disk	1.0
1-in. Disk	1.4
$1^{1}/_{2}$-in. Disk	2.0
$1^{1}/_{2}$-in. Turbine	4.0
2-in. Compound	2.5
2-in. Turbine	5.0
3-in. Compound	16.0
3-in. Turbine	28.0
4-in. Compound	25.0
4-in. Turbine	75.0
6-in. Compound	50.0
6-in. Turbine	187.5
8-in. Compound	80.0
8-in. Turbine	250.0
10-in. Compound	115.0
10-in. Turbine	400.0

[a] This is determined by the ratio of continuous duty maximum flow rates of meter size and type in gallons per minute to maximum flow rate for $5/8$-in. meters, according to the American Water Works Association, *AWWA C700-C703* (1973).

References and Selected Bibliography

Aaron, H. 1975. *Who Pays the Property Tax?,* Washington, DC: The Brookings Institute.

American Water Works Association. 1991. Water Rates, Manual AWWA M1, Denver, CO: American Water Works Association.

American Water Works Association. 1975. Sizing Water Service Lines and Meters, Manual AWWA M22, Denver, CO: American Water Works Association.

American Water Works Association. 1986. Water Rates and Related Charges, Manual AWWA M26, Denver, CO: American Water Works Association.

American Water Works Association. 1986. Alternative Rates, Manual AWWA M34, Denver, CO: American Water Works Association.

Anderson, M. L. 1973. Community improvements and services costs, *J. Urban Planning Develop.,* 99: 77–92.

Armstrong, C. J. 1981. Fees and charges: an underutilized source of revenue for local government, *Virginia Town and City,* 11: 12–13.

Association of Bay Area Governments. 1984. *Development Fees in the San Francisco Bay Area: A Survey,* Berkeley, CA: Association of Bay Area Governments.

Beatley, T. 1988. Ethical issues in the use of impact fees to finance community growth, in *Development Impact Fees,* Arthur C. Nelson, Ed., Chicago: American Planning Association.

Lillydahl, J. H., A. C. Nelson, T. V. Ramis, A. Rivasplata, and S. R. Schell. 1988. The need for a standard state impact fee enabling act, in *Development Impact Fees,* Arthur C. Nelson, Ed., Chicago: American Planning Association.

Landis, J. D. 1986. Land regulation, market structure, and housing price inflation: lessons from three California cities, *J. Am. Planning Assoc.,* 52, 1: 9–21.

League of Oregon Cities and the University of Oregon, Bureau of Governments Research and Service. 1975. *Systems Development Charges: Financing Service Extensions,* Management Information Service Report, Vol. 7, No. 7, Washington, D.C.: International City Management Association.

Lee, D. B. 1988. Evaluation of impact fees against public finance criteria, in *Development Impact Fees,* Arthur C. Nelson, Ed., Chicago: American Planning Association.

Levitt, R. L. and J. J. Kirlin. 1985. *Managing Development Through Public/Private Negotiations,* Washington, DC: Urban Land Institute.

Litvak, Lawrence, and B. Daniels. 1979. *Innovations in Development Finance,* Washington, D.C.: Council of State Planning Agencies.

Metcalf and Eddy, Inc. 1979. Wastewater Engineering, New York: McGraw-Hill.

Mieszkowski, P. 1972. The property tax: an excise tax or a profit tax?, *J. Public Econ.,* 1, 2: 73–96.

Nicholas, J. C. 1988a. *Calculating Proportionate Share Impact Fees,* Chicago: American Planning Association.

Nicholas, J. C. 1988b. Designing proportionate share impact fees, in *The Private Supply of Public Services,* Rachelle Alterman, Ed., New York: New York University Press.

Nicholas, J. C. and A. C. Nelson. 1988. Determining the appropriate development impact fee using the rational nexus test, *J. Am. Planning Assoc.,* 54, 1: 56–66.

Nicholas, J. C., A. C. Nelson, and J. C. Juergensmeyer. 1991. A Practitioners Guide to Development Impact Fees, Chicago: American Planning Association.

Netzer, D. 1988. Exactions in the public finance context, in *The Private Supply of Public Services,* Rachelle Alterman, Ed., New York: New York University Press.

Porter, D. R. and R. B. Peiser. 1984. *Financing Infrastructure to Support Community Growth,* Washington, D.C.: Urban Land Institute.

Porter, D., P. L. Phillips, and C. G. Moore. 1985. *Working with the Community: A Developer's Guide,* Washington, D.C.: Urban Land Institute.

Public Health Service. Undated. Environmental Health Planning Guide, Washington, DC: U.S. Department of Health and Human Services.

Public Health Service. Undated. Environmental Health Practice in Recreational Areas, Washington, DC: U.S. Department of Health and Human Services.

Raftelis, G. A. 1989. The Arthur Young Guide to Water and Wastewater Finance and Pricing, Chelsea, MI: Lewis Publishers.

Raftelis, G. A. 1993. Comprehensive Guide to Water and Wastewater Finance and Pricing, 2nd ed., Chelsea, MI: Lewis Publishers.

Rhodes, R. 1975. Impact fees: the cost benefit dilemma in Florida, *Land Use Law Zoning Dig.,* 27: 7.

Rose, L. A. 1973. The development value tax, *Urban Studies,* 10 (1973): 271–276.

Rose, L. A. 1976. The development value tax: reply, *Urban Studies,* 13: 71–73.

Sandler, R. D. and E. T. Denham. 1986. Transportation Impact Fees: The Florida Experience, Paper presented to the 1986 meeting of the Transportation Research Board, Washington, DC (January).

Schechter, B. 1976. Taxes on land development: an economic analysis, in *Economic Issues in Metropolitan Growth,* P. R. Portney, Ed., Baltimore: Johns Hopkins University Press.

Schell, S. R. and T. V. Ramis. 1987. Systems Development Charges in Oregon. *J. Am. Planning Assoc.* Development Impact Fee Symposium, 1987 Conference of the American Planning Association, New York.

Scott, R., D. J. Brower, and D. Miner. 1975. *Management and Control of Growth,* Three volumes, Washington, DC: Urban Land Institute.

Index

A

Abandoned projects, credit for, 50
Administrative considerations, of system
      development charges, 45–52
    implementation issues, 47–52
      administrative and accounting
        requirements, 48–49
      annual report, 49
      appeals and individual fee assess-
        ment, 48
      construction credits, 50
      exemptions, 51
      interjurisdictional agreements,
        51–52
      management structure, 52
      refunds, 49
      timing of assessment and collection,
        47–48
    procedural requirements, 45–47
Advisory Committee, 45
Alternative wastewater SDC methods, 73
Anti-tax sentiments, 7
Assessed value growth, projecting annual,
    154
Assessment, 83
Assessment districts, 27
Average cost method, 71, 79, 80
Average population, 141

B

Banberry Court, 18
Banberry factors, 17, 81
benefit districts, 27
border effects, 55

building permit, 47, 48
build-out period, 75
buy in fees, 32

C

Capacity expansion, 75
Capital
    cost, 25
    expansion component, 15
    financing plan, 17
    improvements, 6
      element (CIE), 33, 34, 40, 52
      fee-related, 40
      program (CIP), 22, 33, 74, 112
Certainty of benefits, 21
Certificate of occupancy, 47
CIE, *see* capital improvements element
CIP, *see* capital improvements program
Community
    development in, 20
    welfare, 57
Constitutional tests, 14
Construction credits, 50
Consumption characteristics, local, 119
Cost
    allocation method, growth-related, 75
    identification of, 17
Credit issues, 113
Current local asset value, 100

D

Debt
    issuance costs, 98
    outstanding water/wastewater, 111